COSMOLOGY ON TRIAL

CRACKING THE COSMIC CODE

Pierre St. Clair

TORCHLIGHT SCIENCE
San Diego, California

Copyright © Pierre St. Clair 2011-14
All Rights Reserved
No part of this book may be reproduced in any form or by any means including information storage and retrieval systems, except in the case of brief quotations, without prior written consent from the author

Cover design by Robert Wintermute
Book layout by Natasa Marovic

Published simultaneously in The United States of America and Canada by Torchlight Science, San Diego

Library of Congress Cataloging-in-Publication Data available from Publisher

For more information contact the Publisher:
Torchlight Science
A division of Torchlight Publishing Inc.
email: torchlightpublishing@yahoo.com
www.torchlight.com

Reviews

In the course of his dissertation, St. Clair takes on much of modern science and its concomitant contemporary philosophy, showing where they are right, but, more importantly, where they are wrong. His approach is balanced, enabling the attentive reader to walk away with a true education. *S. J. Rosen – New York*

This book simply makes the very strong point that our cosmology of yesteryear was not based on solid evidence, and our cosmology of today isn't either. *V. DiCara – Japan*

An incredibly well researched book. The author has certainly given one plenty of scope to question previously taught theories showing clearly that science is an ongoing changing subject, fraught with those anxious to prove their latest theories, even with very little evidence to back them. *G. A. du Buisson (Mrs.) – South Africa*

If you are interested in coming to grips with just what science knows and does not know about the universe then this book is for you. *A. Brennan – Australia*

A timely reminder that the modern sense of skepticism engendered by the physical sciences is best directed towards the sciences themselves. *S. Sen – India*

The true beauty of this book is its elegance and simplicity, which is what physicists look for in their equations. *J. Rotman – San Diego*

Scientific progress and complacency go ill together, and St Clair has done a superb job in actually advancing the cause of real science. *M. Lyons – Texas*

A good job of dismantling the modern scientific accounts of the origin of the universe. Well written. *M. Cremo – Los Angeles*

A solid and broad foundation for objective discussion, wherever we stand. *David H. Aycrigg – New Zealand*

Contents

    Why I Wrote This Book 1

    Why You Should Read This Book 6

1 Creation Myths 14

2 The Universe in 1980 30

3 The Laws of Nature 45

4 On the Witness Stand 65

5 The Grand Fallacy 79

6 A Theory of Everything? 96

7 Choose Your Universe 115

8 Fuzzy Math and Faulty Theories 146

9 Quantum Theory 171

10 Parallel Universes 190

11 The Santa Claus Culture 219

12 Hubble's Dilemma 232

13 Redshift Resistance 253

14 Cosmology in Conflict 272

15 The Jury Deliberates 292

16 The Verdict 302

17 For What It's Worth 310

    Acknowledgements 327

    About the Author 329

Dedication

What is right is not always popular, and what is popular is not always right – Albert Einstein

To all the men and women who fought for the truth and to those who are fighting for the truth today

Why I Wrote This Book

When I was a kid I asked my mother, "Where did I come from?" She tried to explain how a man and woman have sex. It sounded really strange.

"Why would anybody do that?" I asked. For parents it's a difficult question to answer. It's easy to understand from an adult perspective yet incomprehensible from a child's perspective. They either accept it as it's told, or think it sounds crazy.

Children are so curious. Soon after, they ask about creation in a broader sense. "Where does everything come from?" This question is given an effortless answer: "Everything evolved over a long time." Or, "Everything comes from God."

Despite best intentions, parents and religious leaders often fail to answer important questions. Many kids are left unconvinced.

As teenagers, we question the answers. We want to know why bad things happen to good people, or why good things happen to bad people. During war, mothers on both sides pray to God that their sons will come home. On both sides sons don't return.

Young people think for themselves. My sister didn't buy that we came from God. She became an atheist. I didn't buy it either, because I thought both atheists and theists were fanatics since there is no evidence either way. So I became agnostic.

I had real questions about the meaning of life. Why are we all here and what's the purpose of everything? To satisfy my intellectual and trustworthy inclinations I needed comprehensive answers.

In 1980, I was fascinated by the *Cosmos* series on TV. The renowned American astronomer Carl Sagan took us on a journey through the universe to discover its secrets. I liked him because he was reasonable, logical, and confident. His explanations were systematic, plausible, and provable. That's how he portrayed the modern scientific understanding of the universe.

COSMOLOGY TODAY

My visit to London in April, 2011, was a turning point in my search for answers. I was staying with friends, who, in the best tradition of British hospitality, offered tea and pleasant conversation on arrival. Seeing I was tired, they showed me where I could rest after a long flight.

The room had a dark wooden shelf running the entire height and length of the wall filled with books; a veritable library. On a whim, I looked over the titles. One title caught my attention: *The Goldilocks Enigma*.

Perhaps the word "enigma" spoke to my own quest for answers. As I browsed through the pages, the author, British physicist Dr. Paul Davies, revealed to readers an issue that has baffled cosmologists since the 1960s.

Science has discovered that the universe is so finely-tuned in every detail that even the slightest adjustments in the laws of nature would mean that life, as we know it, could never have arisen. In other words, everything in the cosmos is "just right" to support life; similar to what Goldilocks discovered at the house of the three bears.

Wow! Why is our universe so precisely tailor-made for life

to emerge? I stayed up most of the night reading. My desire to know was re-ignited. I had a renewed appreciation for the power of science to provide insights and explanations. My thoughts returned to my own adventure to unravel the mysteries of life. I was excited.

My first inclination was to watch the *Cosmos* series again. I got the DVDs and traveled back through time and space with Professor Sagan a second time. After reading *The Goldilocks Enigma* and watching *Cosmos*, a lot of questions arose.

I embarked on a journey to have a good look at the scientific data explaining reality and the origin of the universe. Finding illogical and outlandish assertions in every creation account I studied, I needed to get real facts.

Every civilization and culture since time immemorial has a creation story. Curiously, we see them all as myths—except our own. Ideas about who we are and where we come from are important; they influence our actions. We use our understanding of the past to make decisions about the present to create the future we desire tomorrow.

Ancient cultures always believed their creation story to be correct, and we are equally confident that our scientific story is the most correct. In any tradition, your own personal belief system always sounds best.

Today we have access to all the creation stories that came before us. The general public likes stories with large brush strokes and bold characters. The scientific creation account taught at schools and colleges is also a story—a story based on science—but a story nonetheless.

Most physicists readily admit that what we know today is likely to be revised significantly in the future. Society, however, doesn't thrive on uncertainty. When the scientific creation account is told in the form of a narrative, or story, it filters down to schools and the general public as a concrete, coherent plot. Most, if not all, of the uncertainty is removed. Thus, the creation account is presented as facts to be learned, rather than ideas to be explored.

Contrary to popular belief, cosmology and certainty are presently not in sync. Nor are they uniform across the domain of our universe. Not all scientific explanations are equally valid. Explanations that lack certainty, but are presented as almost certain, border on misrepresentation.

Our world of everyday experience is well understood. The subatomic world is far less known. Our understanding of massive black holes, far-flung galaxies, quasars, and the ultimate origin of matter and energy is even less understood. Explanations pertaining to these lack data and certifiable experiments. They are mostly imaginative interpretations of data that lack the certainty of the everyday experience, in which science provides us with stable technology.

Modern Myths

Science is a superb undertaking. Technology like smart phones and laptops work wonders, or if they don't we can find out why. Science has many benefits and is on solid ground here. But I want to know if the Big Bang story of the universe actually represents reality, or it's just a well-dressed "We don't know."

If the creation story in the Bible is a myth, as many scientists argue, then I wondered to what extent the scientific account might also be a creation myth. *Webster's Online Dictionary* provides the definition of a myth:

> "A story of great but unknown age which originally embodied a belief regarding some fact or phenomenon of experience, and in which often the forces of nature and of the soul are personified."

Why I Wrote This Book

A creation myth is a narrative of how *everything* began and how living beings first came to inhabit the world. In the society in which it is told, a creation myth is usually regarded as conveying profound truths. It is understood as a true story although its actual existence is not verifiable.

To a person outside that society, the same myth has the popular meaning of a legend that may not be factual. In other words, people who aren't taught the story from a young age often do not accept it. That means the myth story is not intuitively apparent, it has to be learned. From learning, there is transmission, and through transmission, different versions arise.

Contradictory versions of creation are not limited to traditional storytellers. There are various versions of the science creation story known only to insiders—the physicists who are trying to unravel the anomalies and the mysteries of the universe. Such disagreements are kept from the public who thrive on certainty. But at what cost?

Science has greatly increased our capability to use and exploit resources, and to destroy traditional myths. Yet many people still accept the moral values of religion long after they give up believing the worldviews they represent. Does science provide us with the values and personal meaning inherent in many traditional explanations?

Why You Should Read This Book

Scientific accounts are assumed to be based on observation and evidence. The creation account of cosmology is unique because authority is not based on objective verifiable data.

Astrophysicists portray their cosmological ideas as scientific (supported by evidence) when they are mostly a cloud of probabilities and caveats. The media presents these explanations as factual science, and the unsuspecting public interprets them as "true accounts."

Today, scientists are finding anomalies that don't fit any prevailing theory. An article in the May/2014 issue of *Scientific American,* Supersymmetry and the Crisis in Physics, highlights a calamity of major proportions.

> "Indeed, results from the first run of the LHC [Large Hadron Collider] have ruled out almost all the best-studied versions of supersymmetry. The negative results are beginning to produce if not a full-blown crisis in particle physics, then at least a widespread panic."

The article reveals that theorists working in particle physics admit they're in a state of confusion. Apparently, there is a "widespread panic" that particle physicists might be barking up the wrong tree.

In Canada, scientists from the Perimeter Institute for Theoretical Physics (PI) claim that the present situation within the physics community is ripe with preposterous speculations symptomatic of an invasive crisis. PI director Neil Turok put it this way: "Theoretical physics is at a crossroads right now...In a sense we've entered a very deep crisis."

Dr. Turok claims string theory and the multiverse idea, which some theorists actively promote, are "the ultimate catastrophe." Science periodicals, blog posts, and forums suggest a revision of fundamental principles might be the solution to the present crisis.

Sometimes a spin is put on an ineffective outcome, "Well, we anticipated this result." The tactic is used to give experiments credibility when the early hype about a projected discovery was laid on thick. But when jobs and grant money are on the line finding no good result is considered a nightmare.

We, the people, believe that most scientific accounts are based on observation and evidence. Yet the creation account of cosmology lacks certifiable data. The emphasis is on philosophical ideas, hopes, and aspirations that masquerade as relevant theories. When explanations are divorced from verifiable data they depart from science and the various versions compete for greater elegance like beauty pageant contestants.

This book examines major issues in physics as would be presented in a court of law. Over 64 renowned physicists and mathematicians present their testimony on the scientific basis of cosmology, as if called to the witness stand.

The trial uncovers the startling fact that physicists admit only 4 percent of the matter and energy of the universe is known. A full 96 percent of the matter and energy of the universe remains a dark mystery!

Other issues challenge our perception of physics. There is widespread concern about major theories like the Big Bang, General Relativity, Quantum Mechanics, String Theory, Gravity, redshift, and the speed of light.

The opening chapters compare and contrast the scientific account of the creation of the universe with the traditional creation story. We discover a secret of physics—the remarkable bio-friendly fine-tuning of the entire universe.

In Chapter Four, we call Stephen Hawking to the witness stand. We replay his interview, which was broadcast on "Larry King Live," and review a scientific panel discussing Hawking's book *The Grand Design*.

Chapters Five through Seven, present a detailed analysis of Hawking's "grand design" creation theory. Various scientists challenge and dispute his points. Our attorney identifies and evaluates faults in his arguments.

In Chapter Eight, we look at mathematics in modern physics. Both mathematicians and physicists reveal that many mathematical equations do not represent the real world. We discover that Newton's Laws and Einstein's Relativity Theory do not explain what causes gravitational force, or whether it is inherent within matter.

Chapter Nine continues examining the gravity problem via Quantum Mechanics and String Theory. These theories are also unable to fully grasp gravity—the most common force in nature. We also examine several situations which illustrate that the speed of light may not be the maximum speed limit of the universe, contrary to popular belief.

Chapter Ten reveals the cosmological theory that many universes exist, collectively called the Multiverse. My wife says it sounds like a brand of breakfast cereal, yet reputable physicists believe infinite parallel universes exist where copies of you and me live different lives. It's a belief because there is no data to validate these ideas.

After reading about parallel universes with copies of everyone, Chapter Eleven examines the Santa Claus culture wherein children are deliberately fooled into believing the North Pole fable is real. Why does society pull off an elaborate hoax to fool our kids?

Chapters Twelve and Thirteen examine the dilemma of Edwin Hubble, the great American astronomer who initially proposed an expanding universe based on redshift interpretation. Later he retracted his original view and claimed that redshift should not be used as proof for an expanding universe. Yet his original abandoned hypothesis is still taught as a scientific fact.

In Chapter Fourteen, numerous astrophysicists argue that the Big Bang didn't happen. The Big Bang model of the universe is no longer the accepted theory of creation for many physicists, including Stephen Hawking. In spite of this, it is still taught worldwide as the scientific creation story!

In Chapter Fifteen, the jury deliberates on the arguments presented at trial. After analyzing the issues, Chapter Sixteen presents the verdict of the jury. Based on the evidence, several renowned scientists comment on the decision.

Chapter Seventeen reviews the latest news in science: "Fraud in the Scientific Literature," the "Higgs Boson," and "Ripples from the Big Bang."

Fact or Fancy

The intent of this book is to impartially investigate the scientific creation account. This is necessary because the story has changed considerably over the last hundred years. On the other hand, traditional creation stories have not changed. Most scientists label them as bronze-age superstitions which science has now dismissed with reliable evidence.

As far as Oxford University Professor Richard Dawkins is concerned, he regards religious faith to be, "blind faith in the absence of evidence. I think it's a matter of belief without evidence. As a scientist and as an educator, I like the idea that we believe things because there's evidence."[1]

He argues that truth means evidence. I like that. So let's accept the premise of Dawkins: we rationally believe things to be true because there's evidence. Our goal is to see how closely the traditional and modern creation accounts adhere to this principle. The result of our stringent method will uncover profound secrets.

Given that every scientific paradigm has epistemological reasons for being right, people may ask why does 'change' happen? Why are waves made so the boat keeps getting rocked when everything is apparently fine as it is?

The scientific method is a gradual step-by-step process. Each succeeding step renders certain previous steps invalid—the paradigm shift. The primary theme of this book is to examine the anomalies that keep rocking the boat.

Physics courses do not teach anomalies. But if we confront the anomalous data we must revise our understanding. Today we have to admit that some things we thought we knew (like ninety-six percent of the universe) are actually unknown.

My method is to research the scientific journals, not textbooks, to uncover what's really happening. I expect challenges to the evidence presented. Some people might not like what I uncover. They may object that I don't know anything, that I have an agenda for exposing certain fallacies, or it's an incomplete account. That's why I let the experts do the talking.

So I say to such critics, "Go to the original sources and debate with them."

Until alleged knowledge is exposed as misconception, misinformed ideas evolve into various belief systems. The cosmology of a hundred years ago is no longer fact. Similarly, what we know today could be considered misinformed belief in 2065. People of the 22nd century might laugh at the quaint under-

---

[1] CBC Television Interview - The Hour, 2007

standing we had in the 20th century. This is true from one generation to another.

The famous rock band, The Who, had a Top Ten hit in 1965 with "My Generation." The song speaks to an older generation that can't understand the new generation. Every succeeding generation experiences this—the generation gap. But before long, young people become the old generation and end up in a similar situation as their parents.

Although people want to be in knowledge, ironically they remain in ignorance if they don't accept new information and reject the old. Unless they're willing to update their maps on a regular basis they stay attached to ignorance because in science yesterday's knowledge is today's ignorance.

For example, the facts I wrote on my physics exam in 1969 have been replaced with updated data. If I gave the same answers on today's exam I would fail. So I have a degree in ignorance. That's the situation with cosmology because the "facts" are in constant flux.

Like most persons, physicists are less than saintly in their public service and respect for the truth. So my book examines the case for science exactly as an attorney would do in a court of law.

The evidence must be beyond reasonable doubt in the minds of intelligent people; similar to the rationale for a jury to decide a verdict. After reading this book the uninformed public will no longer be uninformed.

Now I'm not a scientist or an attorney. But I am an investigative journalist with a pinch of intelligence and a dash of common sense who wants to know what is fact and what is fantasy.

I've done tons of research and put it all in perspective. I present the opinions of the respected luminaries of physics: Copernicus, Galileo, Kepler, Newton, Einstein, Edwin Hubble, Fred Hoyle, Carl Sagan, Richard Feynman, Stephen Hawking, Roger Penrose, Paul Davies, Brian Greene, Max Tegmark, Sean

Carroll and other major players.

This book goes to the heart of the major issues in astrophysics. There's enough compelling evidence here to convince even die-hard skeptics. They may disagree with it or debate it; but I have given the necessary facts, the science, the history, and the analysis so you can see it all in perspective and understand it. You may find some confrontation and controversy, but that makes the book more interesting and dramatic.

Reasonable people want to investigate every feature of a controversy and come to their own conclusions. My endeavor is to give all the information you need, right at your fingertips, to be fully informed about the scientific account of the universe. A balanced publication is always a worthy addition to a thoughtful person's library.

I do not present formulas and equations to make the book dry. My purpose is to inform the public not to bore the public. So if you have a penchant for technical info—just Google it.

In an interview with *Discover* magazine, Lewis Thomas revealed a startling fact:

> "Science is founded on uncertainty …We are always, as it turns out, fundamentally wrong.… The only solid piece of scientific truth about which I feel totally confident is that we are profoundly ignorant about nature …It is this sudden confrontation with the depth and scope of ignorance that represents the most significant contribution of twentieth century science to the human intellect."[2]

As an investigative journalist my job is to determine whether the scientific account is truly factual. Therefore, this book presents the evidence that convinced Lewis Thomas of his conclusion, plus a whole lot more since then.

---

[2] Lewis Thomas, "On Science and Certainty", *Discover* Magazine, 1980, p. 58

My *modus operandi* is simply a fact-finding mission. You, the reader, will be on the jury. You will judge the outcome and decide the issues for yourself.

1
Creation Myths

We begin this courtroom drama with a brief comparison of the scientific and traditional creation stories.

Let me confirm that we start with an impartial view regarding both creation accounts; like a jury must do in a court of law.

In the courtroom both sides must get equal billing and a fair hearing. The motive of an attorney—and therefore this book—is to determine that each account is based on evidence and rational thought. Our examination will explore the nature and validity of each account to separate scientific facts from social fiction.

Civilizations of antiquity tried to discover deep meanings concealed within the physical world. The ancients recognized a mysterious nature underlying the universe and this shaped their ritual practices. Creation stories handed down from the remote past were explanations of the hidden truths behind external events. The word "occult" means, "knowledge of hidden truth."

Modern scientific inquiry has replaced ancient occult rituals, yet the goal is the same: to uncover the profound secrets, and hidden reality, underlying the external cosmic manifestation. The word "science" means, "knowledge." Both the modern and traditional accounts have stories that concur with observations of the natural world.

Ancient cultures followed the Ptolemaic calendar accepting Earth as the center of the cosmos and all the heavenly bodies revolving around it. Their creation stories were founded on this geocentric understanding.

Even civilizations that preceded Ptolemy based their astronomy and calendars on the geocentric model. We have the Chinese calendar, the Vedic calendar, the Babylonian calendar, and the Mayan calendar, as examples. Ancient cultures relied on astronomical data for their daily rituals because they were dependent on the changing seasons to maintain their lives.

All creation stories work within a cultural tradition. Therefore, the power and beauty of the insider's perspective may appear strange to an outsider. Every creation myth has this in common.

For scientists, the Genesis story of the Bible doesn't make sense. For evangelical Christians, the science origins story is bizarre. For anthropologists, traditional creation stories are interesting because they reveal clues about ancient cultures. Each group has its own perspective and bias, which must be shed in order to achieve impartiality.

For an impartial examination of the evidence our attorney proposes to briefly recount the traditional Judeo-Christian creation story in a compressed rendition and then contrast that with the modern science story.

Traditional Creation Story

The creation story coming from the tribes of Israel in antiquity tells how the universe was created by an all-powerful God who set the cosmos in motion.

In the very beginning, on that very first day, God separated the light from the dark. Of course, dark cannot remain in the presence of light so it vanished on that first day. Yet, it wasn't until the fourth day that the Sun was created. For us, that's hard to grasp because the Sun's presence and absence separates light and dark here on Earth.

We gauge time according to the Sun as it rises and sets in the twenty-four hour cycle we call a day. Since there was no Sun for the first three days on earth, should we conclude that Genesis is referring to God's days? Of course, a God day is never defined so we have no clue about that.

The biblical tradition explains increasing complexity day by day as various properties of the universe come into being. On the seventh day God finished creating and rested. Humans were also asked to rest on their seventh day. Evidently, God communicated a lot in those days.

The unfolding story is based on trusted sources, as well as observations of the cycles and intricacies of nature. People in the tradition therefore take it as authoritative empirical evidence. It's not really dogma because the story is open to interpretation.

Furthermore, the traditional story has power because it has framed purpose and meaning about the world, along with a system of ethics for moral behavior. The effects have lasted for thousands of years in the Judeo-Christian tradition.

The biblical creation story is universal although its viewpoint places Earth at the center of the cosmos. Consequently, the Sun, the other planets, and the stars revolve around the Earth.

We can never say how much was lost in translation from ancient Hebrew over thousands of years, but it is a complete worldview for people within its culture. From outside the culture, however, it sounds silly. For most modern scientists it's simply absurd.

Every creation story has to deal with the problem of a beginning, and that problem appears whenever a skeptic challenges, "Who created God?" This is the paradox of every creation myth.

SCIENCE CREATION STORY

In the beginning, scientists can't say whether time, space, matter, or energy existed. They have various hypotheses. They're not sure if even nothing existed. If we want to be honest and frank, there's not much we can say. Whatever is said is only conjecture because nothing can be tested. It might have been a primordial chaos but that's speculation, just words really.

Suddenly, 'something' materialized out of whatever it was. Scientists call that 'something' a singularity. It was smaller than an atom but it contained all the matter and energy that was to become our universe. Of course, accepting that everything in the cosmos was crushed into a single point smaller than an atom, we admit, requires a huge leap of faith.

A small seed will evolve into a huge oak tree, but that seed requires an outside agency. It requires soil, water, and sunshine to become a tree. Scientists don't like an outside agency in their creation story. They say that nothing existed prior to the singularity, so they don't accept this analogy.

In any case, the singularity didn't like being crushed. Apparently, it resisted the crunch with a fierce explosion for no reason, or no reason apparent to science.

We call that massive explosion the Big Bang. It created an incredible heat, an amazing 10^{32} degrees, because all the matter and energy of the universe was crammed into that singularity. Some scientists think the temperature was infinitely hot, but either way, at billions upon trillions of degrees, matter and energy were interchangeable.

Instantly, the stuff from the singularity expanded at extremely high velocity far greater than the speed of light. So the universe grew from tinier than an atom to an unknown enormous size in a fraction of a second. The hypothesis of instant expansion, cosmic inflation, defies all known laws of physics.

Scientists opine that it was caused by anti-gravity which

repels instead of attracts. Theorists have a strong belief that in this first instant after the Big Bang nothing was stable. With the extremely high temperature of creation all forces were unified. But as the temperature dropped considerably, the forces separated from each other.

Of course, there's no way to test this experimentally but it's another accepted hypothesis technically referred to as *spontaneous symmetry breaking.*

In this way, energy began to differentiate quickly into several principle constituents: gravity, dark energy, and electromagnetic radiation.

Gravity separated itself from the union of matter and energy immediately. As the temperature continued dropping, the strong nuclear force also left the union. After that, electromagnetism and the weak nuclear forced separated as well. The result was the four independent forces we know today: gravity, electromagnetism, the strong nuclear force, and the weak nuclear force.

All this took place within the first second of the birth of the universe. A few seconds after the Big Bang, the rate of expansion slowed considerably and the universe cooled down to only a few million degrees.

This instantaneous expansion and subsequent cooling allowed six different types of quarks to form. Quarks are considered to be the building blocks of matter. Within these initial seconds all six types of leptons formed. Leptons are the lighter atomic particles like electrons, muons, and neutrinos.

Next, up-quarks and down-quarks formed the heavier particles, protons and neutrons. A proton consists of two up-quarks and one down-quark. The slightly heavier neutron is formed by two down-quarks and one up-quark.

At this point, anti-matter had already disappeared because matter and anti-matter annihilated each other and turned back into energy. Only a small amount of matter remained because there was a little more matter than anti-matter. All the heavier quarks and the heavier leptons have also decayed. Now the universe is hot plasma, with native electrons and protons having electric charges. The entire universe is electrically charged.

The temperature continues dropping until protons can capture electrons to form atoms that are electrically neutral. So the entire material universe, which was crackling with electrical energy, suddenly goes neutral.

Now, protons and neutrons start merging to form the atomic nuclei of hydrogen and helium. As the temperature of the universe keeps falling, stars, planets, and galaxies gradually form over vast time spans.

This is the modern creation story that science teaches today, in a highly condensed form. Naturally, the same problem of a beginning arises when a skeptic asks how the singularity came into being.

From an outsider's point of view the science creation myth sounds bizarre. But for the insiders—cosmologists and physicists—this story is as normal as going to a party.

Of course, the modern creation myth is open to constant revision. It's a flexible story that continues to grow as more data becomes available. It's powerful if you're a scientist who understands the ideas. But for outsiders, it's as preposterous a story as any coming from fiction. The average person can only shrug his shoulders and say, "Whatever."

To sum up, traditional origins stories explain creation by a supernatural being using fantastic powers unknown to us. The Bible tells it like this: In the beginning, God created the heavens and the Earth.

The modern science story describes a spontaneous occurrence that arose from nothing, and from which everything that exists has evolved. Creation came out of nowhere, from nobody. Terry Pratchett depicts how a scientific bible would start: In the beginning there was nothing, which exploded.[3]

Quite noticeable in the modern story is that everything becomes impersonal. Things happen, but people are marginalized.

Human attributes such as moral character, heroism, nurturing, loving relationships, self-sacrifice, and other human qualities are absent. There's no guidance for society, no ideals to strive for, no explanation for dealing with my world right here, right now.

Dark Energy and Dark Matter

Is there more to the modern story? Author Paul Davies explains in *The Goldilocks Enigma* that the expansion of the universe is accelerating faster than before. But scientists don't know what force or energy causes the acceleration, so they call it dark energy.

This dark force makes the universe expand faster and faster over time. "Dark" means it's invisible so it might be impossible to detect. "Energy" means it's not matter, so there is no other option (there's only matter and energy.) Basically, it's called dark energy because it's mysterious.

Whenever cosmologists use the word "dark" it means they have no idea what it is, but they have high hopes they'll identify it soon. "Whatever it is," Davies says, "if you add up the dark energy responsible for making the universe accelerate, you find that it actually represents a total mass that is more than all the matter, visible and dark, put together."

Dark Matter, on the other hand, is no small matter. Its existence is posited by the gravitational influence it exerts on stars and galaxies, and is thought to be a kind of matter that greatly outweighs ordinary stars and galaxies. It's called dark because it neither emits nor reflects light, in other words it's invisible.

---
[3] Terry Prachett, *Lords and Ladies*, New York, Harper Torch, 1996, p. 7

"Surprisingly," Davis adds, "familiar matter such as atoms makes up only about 4 percent of the mass of the universe, almost an afterthought. About 22 percent is in the form of another sort of matter, as yet unidentified, while 74 percent is in the form of dark energy that pervades all of space."[4]

Accordingly, dark matter and dark energy are among the most elusive substances in the universe. To sum up, we actually have an inventory of the universe—but 96 percent of it is a mystery. In fact, of the 4 percent that is normal matter, most of it is in the form of hydrogen and helium. The rest of the matter that science has discovered is only a paltry 0.4 percent.

We have barely a fraction of knowledge about the universe in terms of its constituents, matter and energy. This acknowledgment by Davies was so startling that I checked with NASA to see what they had to say. Their accounting is basically the same.

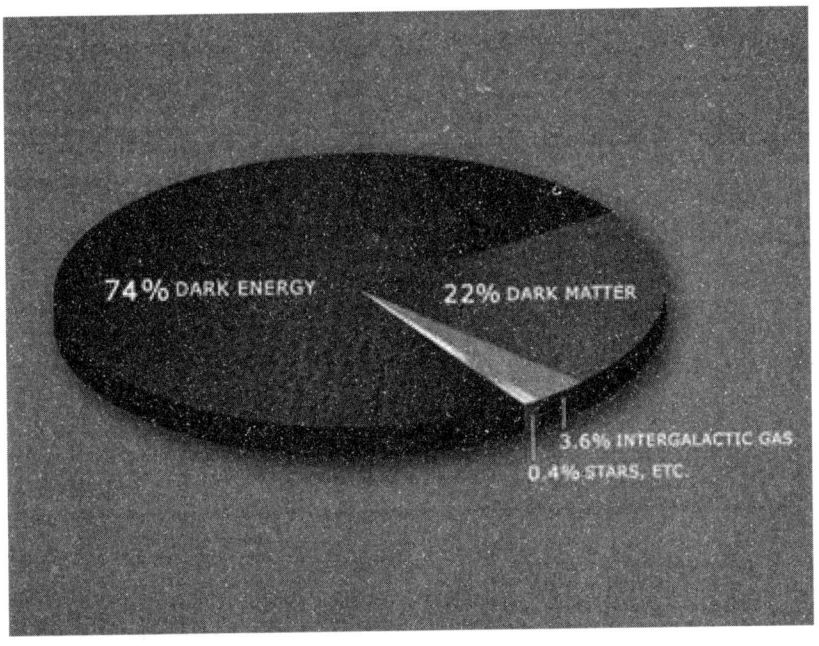

Credit: NASA

---

[4] Paul Davies, *The Goldilocks Enigma*, Penquin Books, London, 2007, p. 139

In spite of all this, on a recent "Larry King Live" broadcast, Professor Stephen Hawking presented a different story. He stated on national TV that, "The scientific account is complete."

Considering all the above, if 96 percent of the universe is unknown to scientists, this begs for an explanation. An unbiased investigator may reasonably ask: what is the actual truth?

Science is on solid ground in many respects, but how can today's cosmology be taught as scientific fact if almost all the matter and energy in the cosmos is a mystery?

Cosmologists readily admit these figures are factually all we really know about the universe. The rest is just theoretical.

We may think we know a lot, but it looks like we understand far less.

THE CASE THUS FAR

For an attorney pleading a case the scientific explanation belongs as much in the realm of faith as the traditional account, because nothing can be tested in either case.

Our attorney has examined the books of various renowned physicists: Stephen Hawking, Paul Davies, Roger Penrose, and Carl Sagan, as well as the seminars, articles, and documentaries by these luminaries of physics.

The idea was to get an overview of what cosmology is telling us about how the universe came into being—the 21st century creation story. We are assured that the modern creation story is based on scientific scholarship, although a lot is left unexplained.

For example, a singularity is not even mathematically describable. The equations have infinite times infinite divided by infinite, and no physicist can work with that. Scientists are trying to describe the origin of the cosmos with something that's not describable by their own methods.

The public is led to believe that scientific evidence is based on employing rigorous canons of academic research, like observation, experimentation, and verification. That's hardly the case with the Big Bang theory.

The history of science illustrates an important lesson. For knowledge to advance there is some risk. We have to discard what was previously considered knowledge. People don't like to be moved out of their comfort zones. They prefer to cling to the old ideas, to maintain the *status quo*.

Another lesson is that all origin stories, other than our own, appear to be lifeless. Look at a butterfly collection. They are beautiful and exotic mounted in an attractive case. But you don't see anything about their life.

Similarly, when I send a postcard of Venice to friends, they see a charming place they'd love to visit. But there's no life, no gondolas in the canals, no fragrance of pizza or pasta. The postcard is only a representation of Venice.

The life of a butterfly loses a lot in the translation to a mounted insect. The experience of Venice is entirely lost in the translation from real city to postcard photo. A representation is like a bridge to the real thing, although it's far from the real thing.

Similarly, the translation of a tribal creation story to a modern audience loses its integral meaning, as does the science story told to a fundamental religious audience.

Broadly speaking, four major elements are found in every creation myth:

1. The story begins with nothing: There's no place, no life, a complete void. Practically speaking, it's a difficult paradox to resolve. Invoking a supreme God is a powerful idea many people can accept. Invoking a singularity that explodes is not as coherent an idea.

2. From nothing, everything comes into existence: Reality starts with simplicity and quickly becomes very intri-

cate. New properties and qualities come into being. An increasing cast of characters and objects acquire the new properties and qualities.

3. The story attempts to explain the how and why of existence: It starts at an accepted given and develops to bring substance and significance to everything. A universal explanation is powerful within a culture.

4. The creation story is satisfying for insiders: However, it makes little sense to outsiders, although it may be entertaining. For them, it's as lifeless as a postcard from Venice; as powerless as a mounted butterfly.

Traditional Creation Myths

Stephen Hawking's book *The Grand Design* portrays a few creation myths. His sixth chapter tells the creation story of the Boshongo people of central Africa and their God Bumba. Then he describes the Mayan origins story.

We may ask: How much is lost in translation from one language to another or from one culture to another? The Boshongo tribal chief might think, *how do I explain it to this English speaking person?* And how does the Boshongo language translate into English? How does a physicist understand the story compared to an anthropologist?

When the manuscript is complete it goes to an editor for editing and a publisher for marketing. Is the retelling of the Boshongo creation story to modern Americans and Europeans an accurate portrayal? Is it possible to understand the Boshongo story from a book the same way as people that live in that culture?

Even more difficult is retelling the Mayan creation story. The Mayan culture was wiped out in the 16th century by the Spanish Conquistadors. From where does Hawking get his information? Perhaps from the diaries of Jesuit priests who accompanied the Conquistadors. But they had a preconceived agenda; convert the savages to Christianity. They claimed Mayan rituals and writings were the work of the devil in order to justify their ethnic cleansing.

Perhaps Hawking salvaged some information from the four surviving Mayan codices, or from artifacts discovered at Mayan ruins by archeologists. Maybe he just Googled online. But before any of this gets to Hawking it has already been interpreted and translated by others. Then he puts his own spin on it because he's an outsider. For him the story is absurd.

Ironically, Hawking's grand design creation story would sound equally absurd to the Boshongo people, the Mayan people, or any culture outside of the scientific tradition. Orthodox Jews and Muslims, for example, might say the modern creation story is ridiculous.

A creation story is appreciated by people within a culture because it relates to their immediate life. Persons outside the culture can never relate to it. For them, it's laughable.

The common problem with all creation stories is that there's little, if any, observational data. So, on what basis can we accept a creation story? One way to test a story is to apply the following parameters:

1. Accept provable data, reject speculation
2. Accept the logical, reject the nonsense
3. Accept the credible, reject the absurd
4. Accept the rational, reject the irrational

At this point in our presentation, the judge calls a recess for lunch. After lunch, our attorney will examine the traditional and modern creation myths keeping in mind these four parameters.

ORIGINS STORIES

The accepted given in traditional origins stories is a supernatural being, infinitely intelligent and powerful, who controls all energy and matter and initiates the creation. The story is accepted on faith and trust in parents and role models. It purports to answer questions, although devoid of hard data.

For the modern story the accepted given is that absolutely nothing existed. From whatever it was, a singularity suddenly appeared to cause an apparent explosion. We accept it on faith and trust from scientists, but again without hard data.

In every creation story the given starting point requires a huge leap of faith. Does one beginning make more sense than the other? Without data, it becomes like a beauty pageant—a matter of personal taste.

Over the past two centuries, the halls of academia favored the modern creation story. As new explanations were added to account for new information, older ones were deleted. The narrative continually evolved.

The modern story is taught in secular institutions from a scientific viewpoint. Evidence is presented that is apparently based on data from observable phenomena that have been tested to support its conclusions. But as we have seen, much of this "evidence" is unaccounted for and based on a leaping antelope of faith. With hindsight the modern story begins to look more like a creation myth.

The traditional story is taught in religious institutions from a perspective of faith. Jewish and Christian schools teach the Genesis story of the Bible. Islamic schools teach the creation story from the Koran.

All creation myths are paradoxical. Some are metaphorical as they try to unravel the universe and give meaning to daily life at the same time. This can be complicated.

Should we look at traditional creation stories from a new

perspective with a more sympathetic eye, and try to enter their meaning, rather than assume they have no value?

There is always more than one point of view. We have the perspective seen by people within a tradition, and the perspective seen by outsiders. There are perspectives within a one-religion country, and within a multi-cultural country. We also have the perspective coming from parents and elders, or from schools and universities.

Accepting a belief system becomes more difficult in a global society where there are many different creation stories. They can't all be true.

Humans need to explain everything in their particular language using words. But it's easier to use symbols that everyone can understand. Therefore most traditional creation stories are told with philosophical and literary sophistication that are, in essence, symbolical. Symbols represent aspects of the mundane world that all insiders of a tradition can understand.

Creation stories are powerful. They provide descriptions of the universe that are meaningful and inspiring. They frame knowledge and guide behavior. But if outsiders don't understand the symbols they can't appreciate the message.

From inside, the traditional creation story is authoritative. Heard from respected elders it has the feeling of truth for young people. There are tenets about procreation and sexuality, about truthfulness and building character. There are details about agriculture and animal husbandry, descriptions of the history of the region, and accounts of human challenges.

Traditional stories contain the trusted information of a society. But there might be two versions of a creation story: one, having a detailed database of empirical facts confirming everything for that society, and another without the boring details.

The modern story, on the other hand, wants to interpret events literally. Therefore, scientists assume that traditional creation myths are taken literally by the elders who tell them, and the youngsters who hear them. This is often not the case.

Let's look at the Genesis story symbolically: *In the beginning* = at the onset of time. *God* = a metaphysical intelligent being. *Created* = formed something from nothing. *The heavens* = space. *And the earth* = matter.

All at once, space, time, matter, energy, came into being from nothing and expanded into everything we see today. The traditional story now appears similar to the modern story taught by mainstream science. The one difference is whether a cosmic consciousness was involved.

All traditional stories invoke a supernatural intelligence of a different nature; a realm of existence above and beyond observable material nature. Science doesn't accept a supernatural existence because it's undetectable with today's technology. But dark matter and dark energy are also undetectable yet scientists do accept that explanation.

In spite of this, when physicists like Stephen Hawking and biologists like Richard Dawkins talk about traditional creation stories they consider them bronze-age myths, or "blind faith in the absence of evidence." They are unsympathetic to values in the narrative. To a caring person these scientists appear judgmental, inconsiderate, unrefined.

Anthropologists are scientists too but they seem a lot more kind, accommodating, and sensitive. Barbara C. Sproul, author of *Primal Myths: Creation Myths Around the World*, explains in her Introduction:

> "Not only are creation myths the most comprehensive of mythic statements, addressing themselves to the widest range of questions of meaning, but they are also the most profound. They deal with first causes, the essences of what their cultures' perceived reality to be. In them, people set forth their primary under-

standing of man and the world, time and space."

Sproul's explanation can also apply to the science creation story because the rational motives are similar—the quest for truth. All origin stories contain a large store of empirical content, which has been carefully analyzed and built upon the knowledge available within a particular society.

Both stories have their trusted authorities as well. The traditional story, however, has the added aspect of exploration into questions of meaning. The science story assures us there is no meaning to the creation.

Personally, I want to remain open to explore the profound significance of reality to my full satisfaction. Endowed with an inquiring mind and having no vested interest I find this prospect both fascinating and exciting.

So as this courtroom drama develops, our attorney will take a closer look at cosmology in today's culture. He will also look at the cosmology of ancient cultures. Adjourn for the day —pick up from here tomorrow.

2

The Universe in 1980

Thirty years ago I was enthralled watching Carl Sagan's *Cosmos* series on TV. Sagan was a likeable guy, good looking and highly intelligent. He was Mr. Personality with a smooth delivery that was convincing. The way he explained science it sounded almost like poetry. He could easily have been the guy who sold ice to the Eskimos.

Yes, it was a magnificent production. Today the DVD set is called *Cosmos Collector's Edition*. In one episode, Sagan visits India. In the courtroom this morning, our attorney wants to replay that segment for the jury.

COSMOLOGY OF ANCIENT INDIA

The segment begins with Sagan pondering India's ancient way of life, which is still alive today. "Cosmology brings us face to face with the deepest mysteries," he remarks, "with questions that were once treated only in religion and myth."

The camera cuts to shots of agrarian India. Village people and their cows are engaged in diverse daily activities. Sagan quotes an ancient text, the Rig Veda, explaining that it was written 3,500 years ago in the ancient Sanskrit tongue. (In my research various scholars date the Rig Veda from 3,500 to 5,000 years old.)

Sagan continues speaking over shots of local customs and rituals.

"The most sophisticated ancient cosmological ideas came from Asia, and particularly from India. Here, there's a tradition of skeptical questioning and un-self-conscious humility before the great cosmic mysteries. Amidst the routine of daily life, the harvesting and winnowing of grain, people all over the world have wondered, where did the universe come from? Asking this question is a hallmark of our species. There's a natural tendency to understand the origin of the cosmos in familiar biological terms."

Of course, the natural tendency to explain the cosmos in familiar language allows all manner of ordinary people to digest the teachings. Using familiar biological terms means facilitating communication of profound mysteries on a level everyone can grasp, i.e. a symbolical or allegorical explanation.

"The Big Bang," Sagan proclaims, "is our modern scientific creation myth. It comes from the same human need to solve the cosmological riddle. Most cultures imagined the world to be only a few hundred human generations old. Hardly anyone guessed that the cosmos might be far older.

"But the ancient Hindus did. They, like every other society, noted and calibrated the cycles in nature: the rising and setting of the sun and stars, the phases of the moon, the passing of the seasons.

"All over South India an age old ceremony takes place every January, a rejoicing in the generosity of nature in the annual harvesting of the crops."

As Sagan speaks the viewer sees everything he describes. Rural India comes alive on the TV screen.

"Every January, nature provides the rice to celebrate Pongal. Even the draft animals are given a day off and garland-

Cosmology on Trial

ed with flowers. Colorful designs are painted on the ground to attract harmony and good fortune for the coming year. However, this is not merely a harvest festival. It has ties to an elegant and much deeper cosmological tradition."

After watching the sights and sounds of South India, the viewer sees Dr. Sagan reappear on camera to pose this question.

"If there are cycles in the years of humans, might there not be cycles in the eons of the gods? The Hindu religion is the only one of the world's great faiths dedicated to the idea that the cosmos itself undergoes an immense, indeed an infinite number of deaths and rebirths."

The above is a reference to the oscillating or cyclic universe model where a Big Bang is followed by a Big Crunch indefinitely. It seems the astronomers of ancient India proposed the model several millennia before modern cosmologists.

The camera now cuts to an ancient temple carved from granite. Professor Sagan walks on set with an enigmatic smile and utters the following remark.

"It is the only religion in which the time scales correspond, no doubt by accident, to those of modern scientific cosmology."

At this point our attorney stops the playback to make his argument. Excuse me? No doubt by accident? What kind of scientific conclusion is that?

The phrase "no doubt" must mean no doubt in Sagan's mind. It can't mean no doubt in other people's minds unless he did a statistical survey. Moreover, saying "no doubt" should mean that data is available to back up a statement. Otherwise it's not science, just a personal opinion.

Furthermore, "no doubt by accident" reveals something deeper—a biased mindset. Could the ancient time scales correspond to modern cosmological data by sheer accident? It's an implausible and unreasonable conclusion.

If they got it right, they got it right. Let them have the credit.

Why diminish it by saying it was by accident, and there's no doubt about it whatsoever?

By declaring "no doubt by accident" Professor Sagan clearly reveals a close-minded mentality in spite of comparable time scales. Even when they get it right, he will deny their success.

Sagan's statement appears to be a Freudian slip, which is always fraught with meaning according to Freud. He established that any slip is actually a sign of a deep emotional truth.

The phrase, "no doubt by accident," ultimately says nothing about ancient Hindu astronomy. Rather, it says a lot about Sagan's subconscious bias.

Suppose a Hindu physicist said this on TV: "Modern physics is the only discipline of all the great works of the West whose astronomical data corresponds, no doubt by accident, to that of the Vedas of ancient India." Wouldn't he be summarily dismissed as quack?

Now that Sagan has passed away we can't ask him why he said "no doubt by accident." But the fact remains, he refused to accept evidence that stared him in the face, even after admitting the ancient Hindu time scales corresponded to those of modern science.

Could an ancient civilization derive time scales that correspond to scientific reckoning without modern technology? Was it really a coincidence? Does no data equal no doubt?

Every human being has four basic shortcomings:

1. We all make mistakes because to err is human.

2. Everybody has imperfect senses. We invent telescopes and microscopes to extend our sense organs. But technology is fallible and the mind that interprets data is also flawed.

3. We have a tendency for cheating; whatever it takes to acquire money, gain recognition, get published, etc.

4. Anyone can be overcome by illusion—that's what a mirage is. Illusion means to accept one thing for another. In the dead of night a piece of rope might be mistaken for a snake... but never during the day.

Even with such shortcomings humans try to accumulate knowledge. Clearly, the ancients had some astronomical ability. To suggest that they simply guessed the size and scale of the universe is not a rational conclusion.

So how did they record their calculations without the technology we have today? I don't know, but it speaks volumes for their intelligence. If their methods revealed figures comparable to modern data, then they deserve to be acknowledged.

Anyway, let's get back to Sagan in India. He continues talking about ancient Hindu cosmological time scales.

"Its cycles run from our ordinary day and night to a day and night of Brahma—8.64 billion years long."

At this point Sagan moves on to another topic and leaves the viewer hanging. My research shows that a day of Brahma, as Sagan noted, is indeed 8.64 billion years. It's called a *kalpa* in the ancient Sanskrit tongue.

I also discovered other interesting facts about ancient Hindu cosmology. I will elaborate on these in Chapter Ten. For now let's continue with the *Cosmos* segment which reveals three things about Professor Sagan.

1. He did research the ancient Sanskrit texts
2. He didn't tell us the whole story
3. He revealed a biased mentality

Cosmology in the Americas

Sagan next turns his attention to other early cultures and their astronomical achievements. The ancient Mayans were such expert astronomers that their knowledge rivals our own. For example, they calculated the orbit of Venus so precisely that they even had a calendar based on its position.

They accurately foretold two profound astronomical alignments. First, on December 21, 2012, the winter solstice Sun aligns with the exact center of our Milky Way galaxy.

Modern astronomers also recognize this event when the sun crosses the plane of the center of the galaxy on that December solstice day. This is a rare cosmic alignment that, according to the Maya, happens only once every 25,625 years.

The second astronomical rarity also occurred in 2012; the earth completed a wobble around its axis. Astronomers call this phenomenon precession. As the earth slowly wobbles on its axis it effectively changes our angular orientation to the larger galactic perspective.

One complete precessional wobble takes about 26,000 years and the exact date of completion was December 21, 2012.

Two galactic alignments happened on exactly the same day, and this very rare event was foretold by the ancient Maya. NASA's own studies do not contradict these ancient predictions.

The Maya attached such importance to this double alignment, and considered the event so extraordinary, that it permeated to the roots of their culture. Can anyone conclude it was "no doubt by accident?"

We may also note here that no "end of the world" prediction was found in the Mayan texts. They simply cited the end of one epoch and the beginning of another. Each epoch was called a *baktun*.

The Mayan astronomers and the astronomers of ancient India clearly understood huge cycles of time. They were also capable of calculating these great cycles with precision; without using modern technology.

All over the world ancient stone monuments were built as calendars to chart the course of the stars and planets. The most well-known sites are Chichen Itza in Mexico, Ankgor Wat in Cambodia, Stonehenge in England, and Abu Simbel in Egypt.

These monuments of human ingenuity reveal the depth of knowledge of the ancient builders. These structures certify the astronomical achievements of the ancients. Why would anyone deny this evidence?

Even in the American southwest, the Anasazi people of New Mexico built a stone temple over one thousand years ago. It was an astronomical observatory designed to mark the day of the summer solstice.

The Anasazi astronomers built their ceremonial temple in such a way that the Sun's rays would shine through a particular window, and light up a particular niche, only on the summer solstice day. This level of precision proves they had the knowledge to track the course of the Sun.

It was important for them to know when the Sun reached its northern limit, the summer solstice day. From that day on, the Sun begins moving south indicating that winter will soon bring cold weather.

This temple had 28 particular niches, which might represent the days of the Moon's monthly cycle. Or, perhaps they mark when the moon reappears in the same constellation. We don't know the details but we do know that these people organized their lives with precision according to the cycles of the Sun and the Moon.

The knowledge of the Anasazi people to track the movements of the heavenly orbs is a triumph of their intelligence. Carl Sagan noted the achievements of the Anasazi astronomers in his *Cosmos* series. Does he give them the credit they deserve?

Once again, he reveals his mindset when he states, "Now some alleged calendrical devices might be due to chance. For example, the accidental alignment of a window and a niche."

How in the world could it be accidental if they accomplished what they set out to achieve? Sagan shows reluctance to recognize achievements for their true worth and significance.

An additional calendar structure sits nearby that not even the skeptical Sagan can dispute. Thus he is forced to admit, "...there is another solstice marker, this one of singular and unmistakable purpose. The deliberate arrangement of three great stone slabs allows a sliver of sunlight to pierce the heart of a carved spiral only at noon on the longest day of the year."

Why is it difficult for Dr. Sagan to acknowledge that ancient cultures could accurately do astronomy without modern technology? He seems inclined to pawn it all off to chance even when faced with the evidence of the great times scales of Vedic India, and the Mayan forecast of the rare double alignment of the earth's precession with the sun's alignment to the center of the Milky Way. And even after personally witnessing the ancient Anasazi ingenuity to build a structure wherein one sliver of light shines at one exact moment in the entire year, the summer solstice at noon.

Today, the parched ruins of the Anasazi have outlived their inhabitants. The Sun Calendar temple has survived the ravages of time. It continues to catch the Sun's rays through that one window to light up a dark niche and mark the summer solstice day.

Apparently, ancient astronomers were not less advanced than their modern counterparts in the ability to understand and predict astronomical events. Yet some modern astronomers seem loathe to admit it.

This apparent prejudice is where our attorney takes issue with modern science. So in this courtroom drama he will take up the challenge of Professor Sagan. The goal is to determine if modern cosmology is beyond suspicion in the nature of its methods.

In 1980 Carl Sagan told us the universe was 15 billion years old. Physics courses teach that the universe is 13.7 billion years old. The figures keep changing from time to time. The latest figure is 13.82 billion years. [bbc.co.uk/news/science-environment-21866464]

Dr. Paul Davies and NASA have already owned up to the fact that 96 percent of the universe remains a mystery comprised by the two unknowns, dark matter and dark energy.

There are other issues in the modern creation story that remain unexplained. Before we analyze these, let's take a brief look at how our conception of the universe evolved over the last twenty centuries.

Cosmology in Ancient Europe

In and around the Mediterranean region 2,000 years ago, the prevailing view of the universe was formulated by the eminent astronomer and astrologer, Claudius Ptolemy of Egypt. He lived near the renowned Library of Alexandria which housed the knowledge of ancient Middle Eastern culture.

Like most people in the region, Ptolemy believed the Earth to be the center of the physical world. His idea of the universe was a geocentric model. Based on simple observations, it's quite normal to believe that the Sun and the planets revolve around the Earth. It's what we see every day.

Ptolemy taught that the heavens were harmonious and changeless. He believed that the motions of the planets through the stars of the zodiac were portents of events on Earth. In those days, astrology was a language that explained various influences on the lives of people. Ptolemy was quite proficient at understanding that language.

Over time, Europe came under the sway of the Roman Catholic Church whose teachings confirmed the Ptolemaic view that Earth was the center of the universe. Humanity was considered the crown of creation.

Supported by the church during the dark ages, Ptolemy's model was the astronomical standard for about 1500 years. However, his model began to lose favor when a Polish cleric, Nikolay Kopernik, introduced a different concept to explain the motion of the planets. Kopernik preferred to use his Latin name and thus he was known then, and today, as Nicolaus Copernicus.

In 1543, the book of Copernicus, *De Revolutionibus Orbium Caelestium Libri VI*, was published posthumously. A different model of the solar system was presented, with the Earth and planets revolving around the Sun. This different model was the heliocentric concept of the world. It demoted the importance of the Earth. The Sun now became the center of the universe.

Copernicus wrote, "Finally we shall place the Sun himself at the center of the Universe. All this is suggested by the systematic procession of events and the harmony of the whole Universe, if only we face the facts, as they say, 'with both eyes open.'"

He noted, however, that despite his heliocentric model, Earth was almost at the center of the universe. "Although [Earth] is not at the center of the world, nevertheless the distance [to that center] is as nothing in particular when compared to that of the fixed stars."

The Copernican model worked at least as well as Ptolemy's model, but his book created an upheaval in medieval Europe. The authority of the Roman Catholic Church held the belief that Earth occupied the central position within the universe. The teachings of Ptolemy, Aristotle, and St. Thomas Aquinas perpetuated this belief. But the church had elevated the idea to dogma.[5]

The church considered the radical concept of Copernicus to

---

[5] Dewey B. Larson, "Globular Clusters", *The Universe in Motion*, North Pacific Publishers, Portland, Oregon, 1984, pp. 33, 37

be a direct challenge to its authority. Thus, his work was put on the list of forbidden books. The magisterial condemnation was based on a condition. If corrections were made to the manuscript, a heliocentric view could be presented as a hypothesis, but not as a fact. This was later reaffirmed in 1620 and remains to this day.

It should be noted here that the heliocentric model of the cosmos was not an original concept of Copernicus. His placing the Sun rather than Earth as a center was not the result of his observations. We associate heliocentric ideas with Copernicus, but he simply reworked ideas he learned from ancient Greek manuscripts, and from the sixth book of Plato's Republic. This is clear by the numerous references and praise for Greek astronomers in his book.

The heliocentric idea was initially proposed by an Ionian philosopher and astronomer who lived sixteen centuries before Copernicus. Aristarchus was able to understand the basic scheme of the solar system and he correctly located our place within the system. He proposed a heliocentric world with all the planets revolving around the Sun. He also postulated that the Earth rotated on an axis once every day.

In ancient India, Aryabhatta was expounding similar ideas. But his Sanskrit works were never translated and remained hidden from Europeans.

The most remarkable consequence of Copernicus was his effect on the future of astronomy. In his own lifetime, the Ptolemaic system reigned supreme. Except for scientific circles, Ptolemy's system satisfied everybody else and this was encouraged by the church. Or, perhaps, the success of Ptolemy over Copernicus was promoted by the church.

Science writer Kitty Ferguson raises an interesting point:

> "Paradoxically, the enormous success of Ptolemaic astronomy is not an argument in its favor.

> It can account for all apparent movement in the heavens. It could also account for a great deal that never happens. It allows for too much. Copernican astronomy, as it has evolved, allows for far less. It's easier to think of something that Copernican theory could not explain. The more scientific way of putting this is that Copernican theory is more easily "falsifiable" than Ptolemy's, easier to disprove. Falsifiability is considered a strength...if new discoveries don't undermine it but fall neatly into place..."[6]

In science, *vox populi* is not always the best qualification to remain the dominant theory. But the idea of Copernicus was the beginning of a new creation story, the science creation story. The modern story evolved, of course, from the sun as center of the universe, to the sun as the center of just our tiny solar system. Ironically, the Copernican idea of the Sun as the center of the universe still put the Earth almost at the center. The net result was tiny as Copernicus himself noted.

Confrontation between the camps of Ptolemy and Copernicus had increased by the time Johannes Kepler appeared on the scene in the late 16th century. Much like Ptolemy he was an astronomer and an astrologer, yet he favored the Copernican view.

Cosmology in Medieval Europe

Young Kepler grew up in a German Protestant monastery. After marriage, he still remained obsessed with God and the creative power of nature. He always desired to be part of God's plan for the world and this inspired all his achievements.

---

[6] Kitty Ferguson, *Measuring the Universe*, New York: Walker and Company, 1999, p. 107

In 1589 Kepler enrolled at a prestigious university where one of his teachers revealed to him the radical ideas of Copernicus. By embracing these ideas, Kepler believed he could understand God's plan for the universe.

During Kepler's time, science didn't have any idea about the physical laws underlying nature. Only the six planets Mercury to Saturn were known during Kepler's life. He wondered why specifically six planets and no more, not realizing that others might just be unobservable. He also wondered about the specific spacing of their orbits.

One day, Kepler was summoned to Austria to assume the position of a mathematics teacher. It was at this school in Graz that he had an epiphany: there were six planets (five plus Earth) because of the Pythagorean idea that only five perfect solids existed from all possible three dimensional shapes.

Kepler believed the five perfect solids held the invisible supports of the spheres of the other planets. He was certain that the connection between geometry and astronomy would reveal the hand of God as the divine mathematician.

"The intense pleasure I received from this discovery can never be told in words," he wrote. "Now I no longer become weary at work. Days and nights are passed in mathematical labor until I can see if my hypothesis will agree with the orbits of Copernicus or if my joy will vanish into the air."

Kepler believed he could glimpse the image of a perfect universe by geometry. But his work with the five perfect solids was ultimately a failure. None of the planetary orbits agreed with each other as he had hoped. Undaunted, he continued pursuing this geometrical illusion. Finally, he came to the conclusion that his observations must be at fault.

The only man in Europe who had access to more precise observations was a wealthy Danish nobleman, Tycho Brahe. He had collected his data over many years and it was the most precise ever recorded at the time.

Brahe had previously written Kepler to join him, and Kepler finally agreed to come after Brahe was appointed the official mathematician of Prague. Thus, Kepler left Graz with his wife and step-daughter. Although the journey to Prague was difficult and tedious, he tolerated all hardships in order to obtain the prized astronomical observations of the great Tycho Brahe.

Unfortunately, the two never hit it off. Brahe was an observational genius and Kepler a great mathematical theorist, but a personality clash prevented their joint work. Within the first year of their meeting, Brahe died. Some say from overindulgence of food and drink. Others say at the hands of Kepler.

In his diary, Kepler reveals that he held Tycho in disdain.

> "Brahe may discourage me from Copernicus (or even from the five perfect solids) but rather I think about striking Tycho himself with a sword...I think thus about Tycho: he abounds in riches, which like most rich people he does not rightly use. Therefore great effort has to be given that we may wrest his riches away from him. We will have to go begging, of course, so that he may sincerely spread his observations around."[7]

After some effort, Kepler finally extracted the prized observations from the reluctant Brahe family. He wrote, "I confess that when Tycho died, I quickly took advantage of the absence, or lack of circumspection, of the heirs, by taking the observations under my care, or perhaps usurping them..."[8]

I noted earlier that physicists are not immune to personal

---

[7] Letter to Michael Maestlin, February 16, 1599, Gesammelte Werke, vol. xiii, p. 289 seq. Partially translated from the Latin by Joshua Gilder and Anne-Lee Gilder, *Heavenly Intrigue: Johannes Kepler, Tycho Brahe, and the Murder Behind One of History's Greatest Scientific Discoveries* (Doubleday, 2004), p. 132

[8] Letter to D. Fabricius, February 1604, Gesammelte Werke, vol. xv, p. 231 seq., cited in *The Sleepwalkers*, Pelican Books Ltd., England, 1959, reprinted 1979, p. 350

motivated agendas, and are hardly saintly in their public service. Now, with Brahe's records in hand, Kepler worked with passionate intensity. After many unsuccessful attempts, he finally discovered that all of Tycho's observations fit by using elliptical orbits for the planets. An ellipse is merely a squashed circle.

Clearly, the secrets of cosmology were always there to be discovered by dedicated stargazers. This knowledge was known by the ancient astronomers of India, the Mayans of Mexico, and the Anasazi of southwest America. But Kepler strengthened the science creation story with his understanding of planetary motion in mathematical terms.

Now, for the first time, European astronomers could forecast a planet's position using Kepler's mathematics. Carl Sagan declared, "Kepler was the first person in the human species to understand correctly and quantitatively how the planets move."

Yet archeology proves Sagan wrong. It's now clear that many ancient cultures accurately recorded and predicted the motion of stars and planets thousands of years before Kepler. Sagan is simply tooting the horn for science.

Court is adjourned for the day. Tomorrow we examine other major issues in cosmology.

3
The Laws of Nature

Day three in the courtroom. Yesterday, we learned that scientists are not always unbiased, and 96 percent of the universe is a huge mystery. Today, our attorney will raise complex issues that science has yet to answer. He begins with an overview of science history.

Every culture has recognized that nature is not an unpredictable muddle. People understood there was an obvious inherent scheme at work and they took advantage of that knowledge to prepare for the future. Utilizing the order in nature they improved life through abundant food production and medicine.

The discovery that the laws which govern nature could be described abstractly using mathematics, improved life through the rise of technology. The question, however, remained: how did these laws come into existence? Was it by accident via undirected random interactions, or was a conscious intelligence behind it all? The Greek terms *cosmos,* order, and *chaos,* disorder, are the basis of two major theories describing the origin and laws of the universe.

No scientific method has been able to prove or disprove the origins of these laws. The only approach is to use logical arguments to favor one or the other conception. Hopefully, the logic will clarify the debate and lead investigators towards the truth.

Traditional stories tell that they were formed by a supernatural creator so the universe could evolve according to a coherent plan guided by the laws. The ancients postulated that a superior intelligence must control these laws that cause matter to always behave in the same predictable fashion. Hence, God the creator must also be God the lawgiver.

These laws aren't random quirks of nature because they keep our planet delicately balanced, which allows for life to evolve. For example, 70 percent of our planet is water. But it's mostly salt water, unfit for drinking or irrigating land for crops, although ideal for preservation. Only fresh water benefits the land and every living organism on the land. How to make use of salty water?

In the remote past a wise sage may have understood nature's water cycle, the ingenious system by which all creatures are provided with usable water. The Sun converts salty ocean water into a gas that rises to form fluffy cumulus clouds where temperatures are well below zero degrees. Wind currents float these clouds, containing tons of water, towards the 30 percent land areas where fresh water is deposited as rain.

When scientists studied this phenomenon, they called it evaporation and condensation. It's a fundamental cyclical force of nature. By this natural law countless tons of water *float in the sky* from one region to another in the form of vapor to keep the Earth finely balanced to support life.

To demonstrate the power of evaporation on a global scale, modern satellite technology records temperatures, pressures, and electric charges. Our eyes in orbit reveal how isolated forces are interlinked to the fluctuating atmosphere in an intricate system.

When ocean water reaches a temperature of 26.1 degrees Celsius, 180 tons of water vapor rises upward every hour.

About one kilometer high the vapor condenses into clouds and releases its heat. This energy can heat the surrounding air by several degrees.

As the air heats up it continues rising which produces powerful vertical winds. The winds drive the clouds up, sometimes to 15 kilometers high. The effect of the earth's rotation below forces the cloud structure to spin. Thunder clouds merge into a vast circle and a hurricane is born. Gigantic storms redistribute heat around the planet and return ocean water to equilibrium by cooling it a few degrees.

So the physical consequence of the power of water evaporation is what sustains life on our planet. Natural laws always work in various situations to enable life to develop. This is a major issue in science today. Where did the laws come from? Why are they so precisely fine-tuned for life?

The science creation story states that a singularity spontaneously arose from nothing, before time, and before space. Some cosmologists believe that the mathematical laws which govern the physical universe existed in a "nothingness" that preceded space and time.

Heinz Pagels expressed it this way: "It would seem that even the void is subject to law, a logic that exists prior to time and space."[9]

The void is a state of no space, and no time, before the Big Bang. So the logic of Pagels only adds another step prior to the event. He offers no rationale for how the laws came to exist in that void, nor an explanation of how laws existed before existence itself came into being. Can science claim there was nothing in the beginning if, indeed, there were laws?

A second hypothesis suggests that the laws arose from the chaos after the Big Bang. They congealed simultaneously with all the matter and energy that they govern. These are the same

---

[9] Heinz Pagels, *Perfect Symmetry*, (Michael Joseph, London, 1985) p. 347

laws of nature we use for the internet, cell phones, GPS, space exploration, satellites circulating the globe, etc.

The problems with the science creation account do not end with the "beginning", however. We are told that after the Big Bang the universe instantly expanded by inflation, a process which defied all known laws of nature. If cosmic inflation defied all known laws we can logically conclude that either, 1) the specific laws that governed inflation remain undiscovered today, or 2) there were no laws controlling the universe at the beginning.

Some physicists take the view that it's pointless to talk about something "before" the Big Bang, because "time" began at the singularity, thus dismissing the whole question.

We might contrast this view with another popular idea—the singularity came from quantum fluctuations of the void. But that opens a can of worms because nothing exists in the void, by definition. It simply raises the question: where did the laws of physics that determine quantum fluctuations come from?

We may also question, what exactly is fluctuating if time, space, and matter do not yet exist?

A third hypothesis states that after inflation the universe cooled. As the temperature dropped considerably, the laws of nature came into being. The universal laws of nature were caused by a variation in temperature? That's a possibility some physicists ask us to accept. Of course, if we say it doesn't make sense we may be labeled ignorant, or incompetent.

A fourth possibility suggests that the laws of physics cannot have existed before creation, therefore, they must have come into being after the Big Bang and evolved over millions of years to what they are today. But this hypothesis would need to specify the mechanism that creates laws which evolve and still consistently produce dependable results as they evolve.

Sir Roger Penrose has initiated a debate among cosmologists with the hypothesis that some patterns found in the cosmic microwave background originate from a continuously collapsing and expanding universe. In other words, there have been many Big Bangs and Big Crunches in the history of the universe.

THE COSMIC CODE

The laws of nature may be considered as descriptive (they describe events) or prescriptive (they determine events). Every civilization has understood that the laws describe how nature works. People studied and analyzed natural phenomena that happened regularly and predictably.

The discovery that the universe operates by strict laws which can be understood by mathematics was a major breakthrough. Sir Isaac Newton formulated mathematical equations that accurately described the outcome of natural events. He even considered mathematical formulas to be the key to a cosmic code that could reveal the fundamental principle of creation.

Modern scientists utilize the language of mathematics to understand the underlying basis of natural phenomena. In Chapter Eight, we will discuss this methodology in detail.

The discovery of a deeper hidden reality at the atomic level has defied all previous concepts of the cosmos. Quantum theorists offer a different perspective. They view the laws as prescriptive, that nature is controlled by the laws. Clearly, the laws can be seen as both prescriptive and descriptive depending on one's perspective.

Of course, accepting that the laws control nature does not bring us closer to answering the fundamental question—how did they come into being?

Answering this question head on, early scientists like Copernicus, Kepler, Galileo, and Newton, regarded their work as necessarily religious in nature. Their goal was to understand how God created the laws that govern the external world in which they lived. They believed science was the path to know the mind of God.

Modern scientists are typically not religious and strongly advocate that the universe has no meaning or purpose behind it. While admitting that there may be an underlying unity within the laws of nature, they nevertheless claim there is no design, no purpose, and no point to it all. They conclude that no designer exists, no teleological principle, and certainly no God.

Teleology

Teleology is a derivation from the Greek word *telos* which means 'end' or 'outcome'. It refers to a cause and effect principle.

In ancient Greece, Plato and his fellow philosophers believed in an abstract realm of idealized forms. It was a perfect domain where all the laws of nature transcended physical reality yet dominated our lowly world of matter. In the physical world only flawed approximations were possible. Our experience, they taught, was just a perverted reflection of the perfect circle, the perfect cube, and the perfect laws of nature.

In the Platonic transcendent realm pure mathematics existed in a perfect state. But where is that abstract ideal realm located? Are there other levels of existence undetected by physics? We explore this idea in Chapter Ten.

The Platonic concept is called the teleological principle. Wikipedia defines it as: "A philosophical account which holds that final causes exist in nature, meaning that design and purpose analogous to that found in human actions are inherent also in the rest of nature."

Advocates of modern science don't assign any value to the teleological principle. Their creation story explains that the universe simply evolved by undirected random chance. In this view, life is seen as simply a fluke that happened in a meaningless cosmos, a one in a billion trillion fluke of nature. The fact that humans have a mind to grasp and understand underlying laws of this fluke universe is another amazing fluke, in and of itself.

This conclusion, says Dr. Paul Davies, "is probably the majority position among scientists."[10]

Is this a reasonable conclusion? As rational animals, humans in every society have been seen to search for meaning behind every phenomenon. If there is no point or meaning to anything, and if life is an accidental fluke, why does every branch of human society search for meaning?

If the universe is meaningless, or without purpose, there should be no governing laws that are fine-tuned for life to arrive and thrive. Yet these predictable and measurable laws do govern the cosmos and have given rise to our human species. We use these laws to our advantage. They enable us to live in a technological world that works wonders.

The fact that bio-friendly universal laws of nature work, and even produce life, indicates that the universe behaves as if there was some underlying purpose. Would a meaningless cosmos mimic a purposeful cosmos?

Search for Meaning

The search for meaning behind natural phenomena has enabled humans to not only observe the functioning universe but to comprehend some of its underlying principles. We interpret the cosmic code of mathematical equations and we understand relationships between matter, energy, and motion.

---

[10] Paul Davies, *The Goldilocks Enigma*, Penquin Books, London, 2007, p. 295

As Davies so eloquently puts it, "The evolving cosmos has spawned beings who are able not merely to watch the show, but to unravel the plot."[11]

All living creatures on the planet observe the workings of nature, yet only humans can explain them. By such understanding, our insatiable curiosity is quenched. When we harness nature's laws for the welfare of humanity we enjoy the benefits of such understanding. We are now privy to the deeper workings of nature, so useful for our own purposes, yet science dismisses the possibility of purpose in nature itself.

The only other alternative is the random chance creation of the cosmos. Our existence is a rare stroke of luck, an accidental fluke. But that's analogous to appealing to a miracle as an explanation. Are we talking science or religion? The boundaries are becoming blurred...

When a majority of scientists say, "It was just a random accident, end of story," it seems to match the miracle rationale you hear in church. "It was just created by God, end of story." Ultimately, the scientist's explanation is a major flaw in the modern creation story. We will return to this issue in Chapter Seven.

Even though no data exists to support the fluke hypothesis, modern physics and cosmology reject teleology because of the possible implications of allowing the guiding hand of God into the universe. However, the anti-God dogma of science seems no better than the pro-God dogma of religion because both are biased assessments of reality.

Professor Paul Davies, an eminent and respected physicist, explains that the mindset of modern cosmology is hardly unbiased.

> "Many scientists who are struggling to construct a fully comprehensive theory of the physical universe openly admit that part of the motiva-

---

[11] Paul Davies, *The Goldilocks Enigma*, Penquin Books, London, 2007, p. 5

tion is to finally get rid of God, whom they view as a dangerous and infantile delusion. And not only God, but any vestige of God-talk, such as 'meaning' or 'purpose' or 'design' in nature. These scientists see religion as so fraudulent and sinister that nothing less than total theological cleansing will do. They concede no middle ground, and regard science and religion as two implacably opposed world views."

Is this a minority opinion? Hardly. It's the view of the vast majority of scientists in the 21st century as revealed by an insider.

"But will God go quietly?" asks Davies. He contrasts the views of fundamentalist religionists with scholarly theologians.

"The God of scholarly theology," he explains, "is cast in the role of a wise Cosmic Architect whose existence is manifested through the rational order of the cosmos, an order that is in fact *revealed* by science. That sort of God is largely immune from scientific attack." [his emphasis]

Even so, science condemns the teleology concept. Physicists have no definitive answer for the issue of how the laws came into being, why they work, and where they exist. It's accepted as given. There are laws and that's it; don't ask why. Clearly, dogmatic science behaves very much like dogmatic religion.

The opposing views of the traditional origin story and the modern story have formed a debate which continues into the 21st century. Was the universe created by God, or did it create itself, manifesting spontaneously out of nothing or from a singularity?

If creation was nothing more than the result of blind chance and a fluke accident shouldn't our universe be completely unpredictable, random, and chaotic? Why do we live in a precisely ordered world fine-tuned for life?

Religious skeptics challenge, how did God come into being?

The traditional story asserts that God is an eternal supernatural being, without beginning or end. Clearly, an eternal being always exists so the question is irrelevant.

Science has yet to discover a supernatural category of existence; a fact which spawns their doubts. In time, however, a super-nature might be discovered just as science discovered dark matter and dark energy in the 20th century. Both continue to be mysterious and less understood in the 21st century. Nonetheless, they provide explanations for what we see, much like the supernatural explanation.

In contrast with an eternal principle, the singularity has the same physical nature as the cosmos; it had a beginning. So the question remains how the singularity came into being before creation, before existence itself? This anomaly needs to be addressed because it seems to plague the science creation story.

Does Einstein's General Relativity theory resolve this dilemma?

Relativity

Before Copernicus, people looked up at the sky and thought the Sun went round the Earth. Today any school boy can tell those people that the Earth goes round the Sun. It doesn't take a PhD to know that.

Some people ask, "What would it look like if the Sun *did* go round the Earth?" Of course, it would look exactly the same. It makes very little difference which model of the universe we use for calculations, heliocentric or geocentric. The results are equivalent because of Einstein's concept of relativity.

The eminent British astronomer Sir Fred Hoyle establishes this logically.

> "Let it be understood at the outset that it makes no difference, from the point of view of describing planetary motion, whether we take the

> Earth or the Sun as the center of the solar system. Since the issue is one of relative motion only, there are infinitely many exactly equivalent descriptions referred to different centers—in principle any point will do, the Moon, Jupiter....So the passions loosed on the world by the publication of Copernicus' book, *De Revolutionibus Orbium Caelestium Libri VI*, were logically irrelevant..."[12]

The Dutch physicist and contemporary of Einstein, Willem de Sitter, admits much the same: "The difference between the system of Ptolemy and that of Copernicus is a purely formal one, a difference of interpretation only."[13]

It's important to understand the impact of the Copernican model within the scientific community. It means that human beings are no longer at the center of the universe. Even more crucial is to recognize the philosophical implications for society.

Jonathan Katz explains the spin put on this idea: "In modern times this has been elevated into the Cosmological Principle which states that, if averaged over a sufficiently large region, the properties of the universe are the same everywhere; our neighborhood is completely ordinary and unremarkable. We are not special, and our home is not special, either. This is one of the foundations of nearly all modern cosmologies."[14]

So we are merely spectators watching the game from the cheap seats. However, humans are the only spectators in the universe as far as we know. Doesn't that make us at least a little bit special?

A major problem with the Cosmological Principle is the fol-

---

[12] Fred Hoyle, *Nicolaus Copernicus*, New York: Harper and Row, 1973, p. 1
[13] Willem de Sitter, *Kosmos*, Cambridge, Harvard University Press, 1932, p. 17
[14] Jonathan Katz, The Biggest Bangs: The Mystery of Gamma-Ray Bursts, The Most Violent Explosions in the Universe, Oxford University Press, 2002, p. 82

lowing: Physicists discover natural laws on Earth and test them with experiments that are carefully controlled. This allows us to know as precisely as possible which influences are caused by which components of any system. However, we cannot apply such methodology to the entire universe, thus reducing the Cosmological Principle to mere conjecture and not a principle at all.

The modern origins story emerged and diverged from the traditional story after Copernicus. The difference was merely a central Sun instead of a central Earth, not much of a difference considering the vastness of space.

Later, science proposed the theory that the universe always existed, there was no beginning. Of course, this was just an arbitrary addition to the story because there was no observational data to back it up. But it marked a further departure from the traditional story which told about a beginning.

Moreover, from the late 16th to the late 19th century science taught that our Milky Way galaxy was the entire universe. This was the scientific teaching for 300 years. In the early 20th century, however, the modern science story began to change dramatically. This was mainly due to the work of Albert Einstein, Georges LeMaître, and Edwin Hubble.

Before Einstein, the universe was considered to be eternal and to exist in a static state. However, as a result of his work this view began to change. Einstein published his special theory of relativity in 1905. Then in 1915 he published his general theory of relativity. LeMaître worked on Einstein's equations and showed that the universe had a beginning. It wasn't eternal.

In the 20th century powerful telescopes were developed that could peer into deep space and reveal distant galaxies populated with hundreds of billions of stars. The universe was no longer just the Milky Way galaxy. It was composed of billions of such galaxies. Our universe had expanded, and the Sun was now quite ordinary.

By using the new giant telescope at Mount Wilson to explore deep space, Edwin Hubble noted the astonishing phenomenon that the galaxies appeared to be receding from us at an accelerating rate. This implied that galaxies were much closer together in the past. Ultimately, if we continue going back in time, they would be more and more on top of each other. Was this observational proof that the universe might have a beginning?

Einstein did not accept an expanding or contracting universe. In his mind the universe was static. So he added a cosmological constant to his equations to keep the universe from expanding or contracting. He fudged his formula to oblige his opinion.

As physicists continued studying the cosmos it became more apparent that universal laws were delicately balanced and fine-tuned in favor of supporting life. The universe was bio-friendly! Then a more startling realization emerged—if the universe was ever so slightly different, life as we know it would not exist, nor would the universe itself exist.

The British cosmologist Brandon Carter was the first person to write that everything in the universe was just right for life to thrive. Carter published a paper in the 1960s demonstrating that the laws of the universe appear to be a grand setup to favor life. He called this bio-friendly fine-tuning, the Anthropic Principle. We also know it as the Goldilocks Enigma.

Carter triggered a revolution in thinking among cosmologists and physicists. He also created a tidal wave of controversy. Many saw the Anthropic Principle as a quasi-religious explanation.

Goldilocks Events

Aristotle believed in "an intelligent natural world that functions according to some deliberate design." Similarly, the Anthropic Principle recognized a delicate balance in the universe which expressly supports life. The laws of nature offered irresistible evidence for purpose or design. They comprised a system that was extremely fine-tuned.

Why do cosmologists have a hard time acknowledging purpose? Let's look at a machine. It has its own purpose; it does what it's designed to do. Bacteria and viruses do what they do. All insects, aquatics, birds, and animals behave like they're supposed to. Every part of their body works in conjunction with the other parts and every part has its purpose. Ecologically, every living organism works in conjunction with other entities, and every part has its purpose.

The planets, stars, and galaxies also do what they do according to laws we have discovered. The universe is a massive machine that functions according to laws that govern all movement. And if all things in the cosmos are bio-friendly it's quite a task to prove there's no purpose or meaning to the universe. Yet, many physicists are loath to accept this at face value.

A strong anthropic principle arose because of the Goldilocks attributes of our planet, the solar system, and the universe. Each link in the evolution of the universe was necessary for our existence, just like certain events on Earth had to occur in order for human kind to develop. It looked like the universe was set up for life to arrive.

All such developments were governed by a fine balance of the fundamental forces of nature. The interplay of these natural laws throughout the universe were "just right" for the development of intelligent life. Therefore, it was called the strong anthropic principle.

The weak anthropic principle is the converse. It refers to the fact that our very existence restricts characteristics and imposes rules upon stellar environments. Everything *must* be just right for life, as we know it, to exist. In other words, due to the fluke of creation, everything just happened to be right for life to come into being. It's a way to "get around" the idea of purpose and design.

Diverse scientists argue that we're still redefining which conditions are favorable for life. What were once considered impossible conditions are now proven incorrect by extremophile bacteria that live in quite inhospitable environments.

Nevertheless, physicists define the habitable zone for life as a narrow region around a star where conditions are optimum for life to develop. It's called the "Goldilocks zone" because the development of intelligent life requires planetary temperatures to be "just right" for liquid water to exist. The Goldilocks zone in our solar system is tiny and includes only one planet, the Earth.

Some scientists talk about the universe being fine-tuned as a result of randomness. They posit that many universes may exist with many different parameters, all of which are unsuitable for life to take place. By random chance, one among billions of universes happens to have all the right conditions. This is the multiverse concept. We'll analyze it in Chapter Ten, keeping our focus for now on the parameters necessary for life.

We learned that Kepler was upset that the planets didn't move in perfect circles. Then he discovered they move in elliptical orbits. Eccentricity measures how close an ellipse is to a circle.

Earth's orbit has 2.0 percent eccentricity. This is fortunate for humanity because it's close to being circular. Elongated orbits with large eccentricities have huge seasonal temperature changes which are not conducive to life. The 2.0 percent figure keeps our climate stable enough to support life.

According to the known laws of gravity, stable elliptical orbits are only possible in three dimensional space. Three dimensional reality is bio-friendly. Circular orbits are unstable in three dimensional space because a small disturbance by the gravity of another planet would pull Earth off a circular orbit.

In more than three dimensions the gravitational force between two bodies decreases more rapidly. Hence, our sun could not remain in a stable state with its internal pressure balancing the pull of gravity. Instead, it would either fall apart or collapse to form a black hole.

The laws of gravity are extremely fine-tuned. If gravitational force was only a little stronger it would pull the Earth off its present orbit and cause it to spiral into the Sun to be burned to ashes. If the gravitational force was just a bit less, Earth would spiral away from the Sun sending us into a deep freeze. The perfect gravitational force keeps our planet in its orbit.

At the atomic scale, the same thing would apply to electrical forces. If the electrical force was only slightly different, then electrons in atoms would either escape from or spiral into the nucleus. In any of the above cases atoms could never exist the way we know them.

The fact that the neutron is 0.1 percent heavier than the proton determines the universal ratio of hydrogen to helium. This 0.1 percent figure also means that the neutron's mass is greater than the combined mass of the proton, electron, and the neutrino, which allows free neutrons to decay.

If neutrons were ever so slightly lighter, they would not decay without an external energy source. Indeed, if protons were only 0.2 percent heavier, they would decay into neutrons and destabilize the atom so that atoms could not exist and thus, no chemistry!

Previously we discussed quarks, which are the building blocks of matter. The summed quark masses that comprise a proton are optimized for the maximum number of stable nuclei to exist. If the sum of the masses of the quarks were changed by as little as 10 percent, far fewer stable atomic nuclei could exist.

The relationship of a star's mass to a planet's distance from it is another example of fine-tuning for life. The mass of a star determines how much energy it gives. If our Sun was 15 percent larger, Earth would be hotter than Venus. If 15 percent smaller, Earth would be colder than Mars. In either case, our planet would be unsuitable for intelligent life as we know it.

Environmental fine-tuning is essential to support life. Life can only arrive and thrive in a habitat that's conducive for living beings to develop. Environmental factors in a cosmic habitat relate to planetary orbit, distance from a star, and the mass of a star. They arise from the "sweet spot" of the surroundings rather than from physical laws.

Some people argue that physical laws shape the good fortune, or lack of it, in the surroundings. In either case, this is environmental fine-tuning. The age of the universe, the Sun, and the Earth are also environmental factors. Three billion years ago our planet might not have been viable to support intelligent life. Now is the optimum time for life on Earth to thrive because environmental conditions make it precisely suitable.

Dark energy is another factor required for life to exist. Physicists don't know what it is but they do know that if the magnitude of dark energy was only slightly larger, it's likely that galaxies would never have formed. That means stars and planets would not have formed, implying no life also. Our very existence depends on this unknown dark energy, whatever it is, being exactly at its present strength and no more.

There is fine-tuning on a galactic level as well. The production of carbon in massive stars requires an exact numerical 'resonance' in the excitation of nuclei for a nuclear reaction to produce carbon. Once this is achieved, all the heavier elements can be produced as a matter of course. Then, supernova explosions disburse all these heavier elements throughout space and to Earth. Fred Hoyle made this discovery in the 1950s.

The laws of the cosmos dictate that remnants of exploding stars re-condense into a new generation of stars. These stars are then surrounded by planets which incorporate the newly formed heavy elements. Without heavy elements like carbon, life would be unable to exist. Clearly, the interplay of the fundamental forces of nature is fine-tuned for life to thrive.

Even the dynamics of massive stars are bio-friendly. They explode precisely in a manner that seeds the universe, and our Earth, with all the heavier elements. That's why Carl Sagan used to say we are "star stuff."

Theoretical physicists examine mathematical models of the universe. They adjust various components of the laws of physics to study the effects of such changes in a scientific manner. It's now clear that almost all of the fundamental constants in nature are fine-tuned for life. If we adjust these constants by even the slightest order of magnitude, the universe would be qualitatively different. Favorable conditions to support life would vanish and none of us would be here.

A change as tiny as 0.5 percent in the strength of the strong nuclear force will destroy almost all the carbon in every star. A change as small as 4.0 percent in the electrical force will destroy all the oxygen in every star. These small adjustments would destroy any possibility for life to survive. Thus, it looks like the strength of the electromagnetic force and the strong nuclear force are ideal for our existence.

Another important discovery is that if the weak nuclear force was just a little weaker, all the hydrogen in the early universe would have turned to helium and no stars would exist. If the weak force was a bit stronger exploding supernovas would not be able to eject their outer envelopes. They couldn't disburse the heavier elements essential to support life throughout interstellar space.

Almost every fundamental constant in the laws of nature is bio-friendly. Even with modest adjustments, our universe would be qualitatively different and unsuitable for life to thrive. A delicate balance allows complex structures to emerge that are capable of supporting life. Very little in physical law

The Laws of Nature

can be revised without destroying the possibility of life as we know it.

In this regard, Stephen Hawking offers us this gem. "Our universe and its laws appear to have a design that both is tailor-made to support us and, if we are to exist, leaves little room for alteration. That is not easily explained, and raises the natural question why it is that way."

It's not easy to conclude the universe was never designed for life. To accept the data at face value, that's easy. To put a spin on data to arrive at a conclusion that contradicts an obvious face value conclusion is not easy.

The inference is inescapable: the universe is completely bio-friendly. This seems to support the traditional story. For the modern story it raises the question, is random chance a plausible explanation for why the universe is delicately balanced to support life? Is fine-tuning really just one fluke after another, *ad infinitum*?

Nobel Prize winner and professor of astronomy at UC, Berkeley, Dr. Aleksey Fillipenko, was recently interviewed on RT (Russia Today) radio. He was asked to explain the origin of the laws of physics.

AF: "What is the origin of the laws of physics? I don't know. That's a question science can't answer. What if the laws of physics have always existed and they give rise to a universe—our universe—and perhaps even multiple universes? That is a possibility, but it is a possibility that is sort of outside the realm of science because we don't know of any way to experimentally or observationally test whether that is a correct hypothesis."

RT: "Let's start from the very beginning—what gave rise to the universe? Why was there a Big Bang?"

AF: "So, what gave rise to the universe is an interesting question. We don't exactly know the answer to that. But we have some ideas..."

Well, ideas are fine, but we want scientific facts. It doesn't bode well for the modern origins story if science only has ideas but no answers based on valid data. That's the same basis as the traditional story.

4
On the Witness Stand

There are many medical practitioners. Among them, some are just quacks. That doesn't mean the entire field of medicine is a hoax. I raise this point because in the medical field if we come across a quack doctor, we want to expose him. It's a public service to protect people from harm or exploitation by quack treatment. This is logical and reasonable.

The situation is analogous in science. If a physicist manufactures a theory which purports to explain everything, then it's natural to examine his claim to ascertain its validity or its quackery. If quack it must be exposed and broadcast to the public. Thus, the uninformed public becomes the informed public. In this way they are protected from being duped.

Our attorney has arranged for a recent "Larry King Live" broadcast to be replayed for the jury to watch on Day Four. Stephen Hawking is introduced as a Professor of Mathematics at the University of Cambridge for 30 years; a theoretical physicist; the 2009 recipient of the Presidential Medal of Freedom; and the co-author of *The Grand Design* with Leonard Mlodinow.

Hawking is a man with outstanding credentials. His book purports to explain why the universe was created. It's the latest explanation of the modern creation story. King gets down to asking questions.

Larry King: "You say that science can explain the universe without the need for a creator. What is that explanation? Why is there something instead of nothing?"

Stephen Hawking: "Gravity and quantum theory cause universes to be created spontaneously out of nothing."

That's his answer, concise and to the point. He states that gravity and quantum theory create universes. Logically, if gravity and quantum theory are the causing agency then they must have existed prior to creation. Of course gravity and quantum theory are surely something, so how did the universe get created from nothing? No clue from Hawking. The contradiction is left hanging.

This answer seems to mimic the traditional story that God created the universe spontaneously out of nothing. Gravity and quantum theory have now replaced the supernatural creator of traditional origins stories. Hawking borrows the traditional answer—that there was a cause, even in the 'nothingness'—but gives the credit to gravity and quantum theory.

I want to know how gravity and quantum theory existed before time and before space. Where was their location? Were they in another dimension beyond the realm of nothing? Hawking avoids the question, "Why is there something rather than nothing?"

King doesn't appear to be satisfied with Hawking's answer. He wants to know how gravity and quantum theory came to be. So he tries again from another perspective.

LK: "You write that because there is a law such as gravity, the universe can and will create itself from nothing. Can you tell me how that law came into existence?"

SH: "Gravity is a consequence of M-theory which is the only possible unified theory. It is like saying, why is 2 + 2 = 4?"

There you have it. You heard the answer directly from Stephen Hawking. Now you know how the law of gravity came into existence. Or do you? I didn't get an answer I could write on a physics exam.

Factually, 2 + 2 = 4 is due to the values we assign to 2 and 4. Apparently, Hawking assigns values to gravity and M-theory to justify his conclusion. But, there was no answer. Hawking simply added a step. Saying "Gravity is a consequence of M-theory" cries out for an explanation. How many people know what M-theory is?

I'm not buying that the law of gravity is a consequence of M-theory if I don't know what the theory is. If it explains the consequence of the law of gravity and the universe, people want to hear that explanation. That's why Larry King is asking. Will the uninformed public think M-theory is Hawking's theory?

A simple Wikipedia search reveals more than the respected physicist told us. It's a recent idea proposed by Edward Witten who states that M-theory is only an untested hypothesis. Clearly, he downplays his own hypothesis.

> "M-theory (and string theory) has been criticized for lacking predictive power or being untestable. Further work continues to find mathematical constructs that join various surrounding theories. However, the tangible success of M-theory can be questioned, given its current incompleteness and limited predictive power."
> [Wikipedia]

"Limited predictive power"—yet somehow Hawking can predict that the whole universe will arise from it? Moreover, I didn't appreciate that Hawking minimized the merit of King's question: "It is like saying why is 2 + 2 = 4?" In other words, the universe was created spontaneously out of nothing so it is what it is.

But, wait a minute. Everything else is what it is, too! What makes this a satisfactory answer? The same logic will allow any religious extremist to give the same smug response: "The universe is a consequence of God. He holds the key to the only possible unified theory. It's like saying why is 2 + 2 = 4? God is what he is. Case closed. End of story."

Anybody can use the same reasoning as Hawking and the public is left in the dark. We are expected to accept on faith. If such an excuse for logic is considered valid for science it should also be allowed for religion. What's good for the goose is good for the gander.

Hawking avoids an answer because he implies it's meaningless to ask this question. It's like asking, why is 2 + 2 = 4? But can you write "Gravity is a consequence of M-theory" on your physics exam? This doesn't satisfy me, what to speak of a physics professor.

Perhaps in his own mind Hawking thinks he's given a definitive reply, but other physicists do not subscribe to such a consequence: "M theory has been criticized for lacking predictive power or being untestable." We have yet to hear anything substantial to validate that science has an answer to the mystery of creation.

King's next question goes right to the point many in the audience probably want to hear.

LK: "Do you believe in God?"

SH: "God may exist, but science can explain the universe without the need for a creator."

With this response, Hawking reveals a lot more about his belief system than he probably intended. His personal belief is that God didn't create the universe because quantum theory and the law of gravity did.

But we need evidence in science, right? No data is presented in this interview. King patiently continues his questions.

LK: "Your book has stirred a lot of controversy. Why do you think people react so strongly to your contention that it is not necessary to invoke God to explain the creation of the universe?"

SH: "Science is increasingly answering questions that used to be the province of religion."

LK: "One of your colleagues out of Cambridge said that science provides us with the narrative as to how existence may happen, but theology addresses the meaning of the narrative. How do you respond to that?"

SH: "The scientific account is complete. Theology is unnecessary."

Clearly, Hawking does not agree with his Cambridge colleague. But in this exchange he also contradicts NASA physicists by claiming that, "the scientific account is complete." Complete means there is nothing more to discover about how the universe was created. Game over; end of story.

We already learned that scientists acknowledge a full 96 percent of the stuff of the universe is unknown. Yet for Stephen Hawking the scientific account is complete when only 4 percent of the matter and energy of the universe is known? A strange sense of completeness, indeed. And when the remaining 96 percent is finally understood, it won't matter one iota?

These contradictions could well be a source of amusement 50 years from now; a joke among cosmologists in the 2060s. And as science advances into the 22nd century Hawking's Grand Design theory might become as comical as Henry Ford's belief that his horseless carriage was the pinnacle of human engineering.

Seeing that he's getting nowhere, King changes the subject to give the interview a human interest angle. Before closing the interview, however, he returns to the topic of the book.

LK: "What do you most hope people take away from your new book *The Grand Design?* What is the most important point in the book?"

SH: "That science can explain the universe, and that we don't need God to explain why there is something rather than nothing, or why the laws of nature are what they are."

"Science can explain the universe," but clearly Hawking has not explained anything. We've yet to hear a definitive explanation. Saying, "We don't need God to explain why there is something rather than nothing," sounds more like a putdown of religion and less like a scientific explanation.

Because we already know that 96 percent of the universe is unknown, Hawking sounds like a fundamentalist scientist basing his convictions on faith, rather than rational explanations. He presents no data, no evidence, no mention of dark matter or dark energy, or that 96 percent of the universe is unknown.

The thought came to me, *is Hawking simply bluffing Larry King?* It was beginning to look like he was borrowing religious doctrine and mocking religion. He was starting to sound like one of those know-it-all guys you meet in college that nobody likes.

Without data to back up his assertion, Hawking comes off like a religious extremist beating the drum for his own religion —Scientism. I still don't have an answer to use for my physics exam.

Even the name of his book *The Grand Design* smacks of the designer that the Intelligent Design movement tries to establish. But now it seems that Hawking wants to be the intelligent designer. He invokes M-theory over and over with no mention of Ed Witten who rightly deserves credit as the originator of M-theory.

SH: "According to M-theory ours is not the only universe. Instead, M-Theory predicts that a great many universes were created out of nothing. Their creation does not require the intervention of some supernatural being or God. Rather, these multiple universes arise naturally from physical law. They are a prediction of science."

Again, he claims predictive power from M-theory. Is he reading from a script or towing the party line? It's the same slogan with zero explanation from the scientific side. How did the physical law come into being first, so that multiple universes can arise naturally from it?

Hawking has again added an unexplained step between nothing and something, betting that King won't notice. It's the old conjurer's trick. Pull a rabbit out of a hat, or a coin from a child's ear, to give the impression it mysteriously appeared from nowhere. It's the illusionist's stock of trade.

He also informs us that M-theory predicts: "a great many universes were created out of nothing." Existence can be derived from nonexistence by M-theory? That's interesting for a theory which has been described as "untestable" and "lacking predictive power."

Of course, prediction means before the fact, not after the fact. Hawking has a system that proves what he wants to establish, and he insists that the theory predicted it. His system predicts the winning horse, but after the race is over.

His point has become clear. Science can do away with God because M-theory does a better job. It creates unlimited universes, rather than a single paltry one. God is redundant for the function of creation, and Hawking has arrogantly become the Donald Trump of science: "God, you're fired!"

By basing his convictions on such shaky ground this interview doesn't bode well for the reputation of the old professor. His arguments, as well as M-theory, are not his original ideas. What we have from his testimony are just speculations and contradictions. We may buy it or not, but don't expect any data from his *Grand Design* book to back it all up. M-theory is not evidence because it's untested. It's merely a hypothesis.

Did you think Hawking was trustworthy in this interview? I didn't get that feeling. Perhaps he's bluffing, or just being sardonic. In any case, if Stephen Hawking represents where cosmology is at today, then actual science has been abandoned. We're being fed the new religion of Scientism.

According to Wikipedia, Scientism has two equally pejorative meanings:

1. To indicate the improper usage of science, or scientific claims, in contexts where science might not apply, such as when the topic is perceived to be beyond the scope of scientific inquiry;
2. There is insufficient empirical evidence to justify scientific conclusions.

At this point it would be prudent to know what other renowned physicists think of M-theory. We don't have to blindly accept the word of one person just because he's a big name in physics. Intelligent people don't accept things blindly, or out of awe of authority. They do the research so they're not dependent on a single person's opinion. Thus, they come to their own independent conclusions by weighing up the pros and cons from all sources, just as in a court of law.

Here is more research on M-theory. I discovered that several luminaries of physics and cosmology do not even consider it a theory.

On a recent radio show, *Unbelievable? with Justin Brierly*, broadcast on Saturday, September 25, 2010, Professor Alister McGrath and renowned physicist, Sir Roger Penrose, discussed Hawking's book. The discussion leads to the following topic: Is Hawking's M-theory good science?

Alister McGrath: "The first thing I want to emphasize is we have to be careful about the provisionality of this. In some of the debates on Hawking's book, you almost get the impression that science now knows the answer to this question and it's called M-theory. But it's not; it's just a staging post along the long road of science as we try to make sense of things. At the moment this looks quite hopeful, but it's clear that further work needs to be done. And the question is where will that take us in future?"

Roger Penrose: "Well, I think it's actually stronger than that. What is referred to as M-theory isn't even a theory. It's a collection of ideas, hopes, aspirations; it's not even a theory. And I think the book is a bit misleading in that respect. It gives you the impression that here is this new theory which is going to explain everything, and it's nothing of the sort. It's not even a theory and it certainly has no observational data."

Physicist Paul Davies agrees with Penrose. He writes, "nobody has yet written down the equations that govern the full M-theory, let alone solved them."[15]

So, M-theory is actually just an idea. It is not based on evidence. Scientific theories, of course, require the support of evidence. At the very least they need a feasible way to gather evidence to either prove or falsify predictions.

The scientific community is cautiously excited about the potential of M-theory, due to its internal elegance and possible explanatory power, but the fact remains that it's not even theoretically testable. There is ongoing debate whether it can even be considered a theory.

Most M-Theory adherents view it as a possible future scientific explanation. Researcher Brian Greene presents an overview of string theory, and its subsidiary M-Theory, in his book *The Elegant Universe*. He clearly states the uncertain status of such theories and explains the implications:

> "No matter how compelling a picture string theory paints, if it does not accurately describe our universe, it will be no more relevant than an elaborate game of *Dungeons and Dragons*."

If evidence is eventually found to validate M-Theory, it will be a breath-taking achievement for modern science. If not, it could turn out to be the most elaborate game of *Dungeons and Dragons* ever played by nerds. The coin is still in the air.

The next topic discussed on the show: Does Hawking's M-theory show that the universe created itself?

---

[15] Paul Davies, *The Goldilocks Enigma*, Penquin Books, London, 2007, p.129

Justin Brierly: "This is one of the interesting aspects of what he appears to be saying, that M-theory shows that the universe can create itself, so to speak, out of nothing. Is that something which science can ever really tell us or be definitive about?"

Roger Penrose: "It certainly is not doing it yet! I think the book suffers rather more strongly than many. I mean it's not an uncommon thing in popular descriptions of science to latch on to some idea, particularly things to do with string theory which have absolutely no support from observation. They're just nice ideas that people try to…"

JB: "But they will always remain theoretical…"

RP: "Always is too strong. It may well be they'll just be refuted, but even that is something which is very far from observational testability. They're hardly science."

Later in the discussion another question brings up the multiverse idea: Is our fine-tuned universe simply one of many in a multiverse?

RP: "Multiverse means different things. There is this 10^{500} different M-theories, [that number is 1 with 500 zeros after it; that's how many different universes are postulated, each with a unique set of laws] and these are different schemes in which you might have constants of nature being different. But not just that, all sorts of things are different; and the argument is of all these different possibilities we live in one where life is possible. On most of them life is not possible. Now that's one type of scheme. They all kind of co-exist but we find ourselves in…"

JB: "…the one that is habitable for human life."

RP: "That's right, and that's what's called the anthropic argument, which has some justification. It's overused, I think. And this is the place where it's overused. It's an excuse for not having a good theory."

Penrose clearly establishes that M-Theory is not yet a theory. It's only a collection of ideas and aspirations for a possible theory. Because it's untested it can only be considered hypothetical at best, just as Ed Witten admitted. Hawking uses M-theory in a way that is unwarranted, as an excuse for not having a good theory in the first place. This is the precise definition of Scientism.

So it's clear that M-theory is not based on any observational evidence. The absence of evidence means blind faith, according to Richard Dawkins.

Secondly, M-theory is an untested and incompletely devised candidate hypothesis. In spite of this, Hawking invokes it as if it was truth beyond doubt, indeed a truth that needs no further explanation or any evidence at all. We have seen this with religious dogma, and now Hawking presents it in the name of Science. Again, this is clearly Scientism.

Neil deGrasse Tyson, director of the Hayden Planetarium at the American Museum of Natural History in New York, is considered by some as the successor to Carl Sagan in popularizing science. Tyson notes: "There's a dimension to news reporting that I think not all journalists have the talent, frankly, to achieve, and that's to digest information, interpret it, and deliver it in such a way that people have a deeper understanding of what's going on."

Journalists can plead ignorance or unfamiliarity with the material—after all, they are not scientists. But what about Hawking?

Let's take a brief look at his career. As a young man he wanted to study physics under Britain's most notable physicist at the time, Sir Fred Hoyle of Cambridge University. But Hoyle was not accepting new students, so Hawking worked under the supervision of one of Hoyle's protégés, Dennis Sciama.

To be awarded a PhD at Cambridge, a student has to write an original thesis which is a substantial contribution to knowledge. This has to be accomplished within a three year period, so it's a heavy requirement.

Hawking couldn't find a suitable subject for his thesis. "I

made a bad start at Cambridge," he explained in a TV documentary. "I had just been diagnosed with ALS, or motor neuron disease, and didn't know if I would live long enough to finish my doctorate, and I was having difficulty finding a problem for my thesis." [Hawking has struggled against Amyotrophic Lateral Sclerosis, a disease which left him almost completely paralyzed]

In an interview, Sciama recalled the situation.

Dennis Sciama: "In Stephen's case, his first couple of years was a bit slow because he was fallow at that time. He couldn't find a good problem and I couldn't find one for him."

With less than one year remaining, Hawking had still not come up with a worthy project. One day he attended a lecture by Roger Penrose, a young luminary in cosmology whose work on a problem connected with Einstein's theory led to an important discovery.

In his seminar, Penrose presented how large stars collapse to become black holes. Large stars collapse when they run out of fuel, but only a gigantic star can collapse into a black hole. The entire star's matter is crushed into an infinitely dense point—a singularity.

At the time physicists believed this could only happen when a star was perfectly symmetrical because it would collapse uniformly in all directions. Could a star be absolutely symmetrical?

The work of Penrose established that if a star was large enough it could become a singularity whatever its shape. This was considered a very important contribution to the theory of relativity.

Hawking was inspired by Penrose and the work he was doing. One day he approached his supervisor with an idea for his thesis.

Dennis Sciama: "Hawking heard a seminar given by Penrose in which he announced his result. A little later Stephen said, 'We can adapt Roger's argument to the whole universe, in a certain sense that the universe is like a big star.' Of course, the universe is expanding, but if in your mind you reverse the sense of time, then the universe is collapsing. It's a bit like collapsing a very large star. Perhaps you can prove that in that collapse you can again achieve a singularity. It would mean that the Big Bang origin of the universe would have to be singular. So that would be a great discovery. So in his last year he proved his first singularity theorem for the universe on the basis of certain very reasonable assumptions; the Big Bang had to be singular."

In his own words, Hawking describes the outcome. "I was awarded my PhD by showing that Einstein's theory of relativity implied that the universe must have begun with a Big Bang. It couldn't have collapsed, bounced, then expanded again."

This became the Penrose Hawking singularity theorem. Penrose developed the theorems to prove that massive stars collapse into a singularity.

Hawking was invited to the Vatican in 1975. The church was happy that science had proved that the universe had a beginning, as stated in the biblical account, rather than the idea that the universe always existed.

"In 1975, I was awarded a medal by the Pope for my part in proving the Big Bang theory."

This would not be Hawking's only visit to Rome.

"I went back to the Vatican in 1981 for a conference on cosmology, this time under a different Pope. He told us that it was fine to study the universe after the Big Bang but that we should not inquire into the Big Bang itself, because that was the moment of creation and the work of God."

Of course, Hawking wasn't a religious man so he didn't follow the advice of the Pope. The honor he received from the Vatican didn't change his view.

"If science and religion were now at one," he remarked,

"perhaps they were still not quite seeing eye to eye."

Years later, Hawking related the story of meeting the Pope to an American audience. He added some humor to the tale.

"I was glad he didn't realize I had presented a paper at the conference suggesting how the universe began. I didn't fancy the thought of being handed over to the Inquisition, like Galileo."

A projector shows a slide of Hawking in jail. The audience laughs as the Pope becomes the butt of Hawking's joke.

Ironically, his work on the Big Bang theory moved science closer to religion. A beginning was the original premise of traditional creation stories. If there was a beginning then the conception that the universe always existed had to be revised. What science had been teaching for centuries was incorrect.

But that's not the end of the tale. We will now examine Hawking's book *The Grand Design* exactly as would be done in a court of law. Will the book resolve the contradictions we heard in his Larry King interview?

5
The Grand Fallacy

Many cosmologists today admit that even the best of their theories are simply hypotheses. Some of these hypotheses are unobservable and untestable. Few are based on objective data the way the scientific method prescribes and the way scientific empiricists work. Several hypotheses are postulated as just being the most probable interpretation according to the specific model that a cosmologist favors.

The Grand Design, a book by Stephen Hawking and Leonard Mlodinow, is a case in point. Since the jury now has an idea of the modern creation story our attorney wants to determine if the book contributes anything substantial.

First Chapter

Well, right on the first page, the authors take a dig at another discipline: "…philosophy is dead. Philosophy has not kept up with modern developments in science, particularly physics."

This condescending statement could be seen as an insult by many eminent philosophers at Cambridge University, who are actually colleagues of Hawking. Are we supposed to accept that these scholars haven't kept abreast of the world of science?

It's disappointing to see the book start on a negative note. Hawking's reputation as a brilliant physicist means he doesn't have to take a dig at others. But many physicists do complain about philosophers and are negative towards them, although physicists use so much philosophy which they fail to articulate properly, as we shall see shortly.

For now, let's look at the definition of philosophy to clarify what Hawking considers to be dead.

Philosophy: the rational investigation of the truths and principles of being, knowledge, or conduct.

Clearly, a valid investigation of truths and principles of knowledge must be rational. Yet according to *The Grand Design*, rational investigation of truth by philosophers has died. Now only physicists can provide the light of knowledge? So that's a challenge; one that we intend to pursue.

Next, a more detailed definition.

Philosophy: the academic discipline concerned with making explicit the nature and significance of ordinary and scientific beliefs, and investigating the intelligibility of concepts by means of rational argument concerning their presuppositions, implications, and interrelationships; in particular, the rational investigation of the nature and structure of reality (metaphysics), the resources and limits of knowledge (epistemology), the principles and import of moral judgment (ethics), and the relationship between language and reality (semantics).

The Grand Design begins with the denigration of philosophy, the discipline that eliminates nonsense from any discussions of reality. Why is that dead? Because, the authors assure us, we can no longer use common sense for coming to conclusions about reality:

> "...common sense is based upon everyday experience, not upon the universe as it is revealed through the marvels of technologies such as

> those that allow us to gaze deep into the atom or back to the early universe."

From the first page the authors establish a negative approach. They elevate their own discipline to a sublime level by discrediting the discipline of philosophy. They consider the rigorous academic discipline of philosophy to be no more than mere everyday common sense. Now we need to look at the definition of common sense.

Common sense: sound practical judgment; native intelligence.

The reader is asked to abandon native intelligence and sound judgment. Apparently, this will facilitate understanding the subject matter of Hawking's book. In order to accept the view of the authors we can't rely on our own judgment and common sense. So on what basis do we accept the veracity of *The Grand Design*? It seems to be on blind faith.

Right away, the opening pages of the book struck me as familiar. Why it's just *The Emperor's New Clothes* in new dress! This well-known tale by Hans Christian Andersen was first published in 1837.

> An Emperor, who only cares about his appearance, hires two expert tailors. They promise him the finest suit of clothes made from a fabric so fine that it is invisible to people who are stupid, incompetent, or unfit for their position. The Emperor can't see the fabric himself, but he pretends that he can rather than appear unfit for his position, or stupid.
>
> When the swindlers finish his new suit, they mime dressing the Emperor. Everyone in his court goes along with the pretense. When the monarch marches in procession before his sub-

> jects in his new clothes, every person continues on with the charade.
>
> Suddenly, a child in the crowd, too young to understand the desirability of keeping up the pretense, blurts out that the Emperor is wearing nothing at all. Then the cry is taken up by others.
>
> The Emperor cringes, but holds himself up proudly and continues the procession under the illusion that anyone who can't see his clothes is either stupid or incompetent.

Dr. Hawking now drops a second bomb on his readers in his opening chapter. Everything in existence is "created out of nothing" according to the book's interpretation and presentation of M-theory. So now there's no need for the "intervention of a supernatural creator or god."

Fortunately, the fallacy of this statement becomes crystal clear if we choose not to abandon intelligent judgment.

Nothing: no thing; not anything; naught; nonexistent.

The contradiction here is that the science creation story tells us that a singularity caused the Big Bang, so that was something that existed previous to creation; hence the universe didn't spring from nothing.

Nothingness has no properties because there is "no thing" existing that might have properties. It's not logical to say that everything in existence came into being from nothingness, from zero. Moreover, nothingness is a state devoid of anything that might even be a cause. So the idea that 'some thing' came from 'no thing' spontaneously is a faulty conclusion.

A plausible and reasonable explanation might be that a singularity appeared from something which was imperceptible, and misconstrued to be nothing. For example, if a singularity emerged from a higher dimension that was undetectable, we would be wrong to conclude it sprang from nothing. That which is imperceptible does not imply its non-existence.

Another example: the mind of Stephen Hawking is unobservable. Should we conclude that his mind doesn't exist? Of course, after reading the first chapter of *The Grand Design* an argument could be made for that conclusion. However, it wouldn't be a philosophically sound, acceptable explanation, even if his book assures us we can't rely on common sense.

Without sound practical judgment and native intelligence no person could ascertain if an argument was fallacious.

Fallacy: a deceptive, misleading, or false notion, or unsound argument.

The first major problem with Hawking's book is that he asks us to abandon reason and logic. He suggests that from the very first page. Only then can readers be led down the garden path to the book's conclusion. And those who do not accept this premise, are they incompetent to understand?

Isn't this the same explanation the swindlers gave in *The Emperor's New Clothes*? They spun cloth so fine that the foolish couldn't see it.

In the absence of data one may base arguments on some theory to prove a point. The Anthropic Principle, on the other hand, is based on hard data—evidence that the universe is perfectly set up for life to arrive and thrive. Everything in the universe is so finely tuned that it's difficult for cosmologists to escape the conclusion of prior advanced intelligence.

Yet Hawking asks us to abandon normal intuitive reasoning. Why? Well, if the universe only mimics the activity of a bio-friendly intelligent system, then common sense reasoning would lead to a wrong conclusion. We ask the jury to consider: would a purposeless universe mimic a purposeful one?

The observable evidence of bio-friendly laws that govern the universe, the mathematical laws that work, the laws of physics, the growing amount of evidence that the universe exists in clear exactitude for biological life, all of this evidence points to a more logical, reasonable, and plausible conclusion than "it just arose from nothing."

Cosmology needs a good explanation to resolve this fine-tuning question. The modern creation story doesn't want to mimic the traditional story, after all.

The Grand Design attempts to respond to this challenge. The authors declare that because our universe is fully suitable for biological life, with humans as the crown of creation, we can come to the following conclusion. "Although we are puny and insignificant on the scale of the cosmos, this makes us in a sense the lords of creation."

Yes you read that right. This statement has no correlation to the fine-tuning of the cosmos, but since Hawking brought it up we can offer a response. I propose that the 100 percent death rate of every living person puts a nail in the coffin of that argument. Moreover, Hawking will soon present an argument that we have no free will. But if nature has 100 percent victory over us, in what way are we "the lords of creation?"

Professor Hawking explains the purpose of his book is to find an ultimate theory of the universe. "We now have a candidate for the ultimate theory of everything, if indeed one exists, called M-theory." His book intends to answer the why of existence rather than the how:

Why is there something rather than nothing?
Why do we exist?
Why this particular set of laws and not others?

Thus far, there is no mention of Ed Witten who formulated M-theory. But let's move on to see if the authors can live up to their boast.

SECOND CHAPTER

Hawking opens the chapter with that tired old idea of mocking other creation stories. This time it's Viking mythology. Clearly, he wants to find a creation myth that appears implausible and use that to show the superiority of his own creation myth.

The two opening chapters have begun by denigrating different perspectives, including philosophy and common sense. Unfortunately, by adopting the approach that only physics is on the right track, Hawking gives science the appearance of religious fundamentalism—only my view is true. If you don't accept my interpretation you must be ignorant or incompetent. It's a strategy. But is it a good strategy?

After a brief overview of the scientific method Hawking concludes: "Today most scientists would say a law of nature is a rule that is based upon an observed regularity and provides predictions that go beyond the immediate situations upon which it is based." OK, we can accept that.

And, "...most laws of nature exist as part of a larger, interconnected system of laws." Fine. No problem so far. Then, "If nature is governed by laws, three questions arise."

1. What is the origin of the laws?
2. Are there exceptions, like miracles?
3. Is there only one set of possible laws?

These are good questions. They need good answers. The answer to the first question, according to Kepler, Galileo, Descartes, and Newton, was that God created the laws. We know from his opening chapter that Hawking believes the laws came into being from nothing, even though he claims the universe was caused by gravity and M-theory, both of which are something.

The answer to question two is considered "a principle that is important throughout this book. A scientific law is not a scientific law if it holds only when some supernatural being decides not to intervene."

It's clear by now that Stephen Hawking has some fixation about traditional creation stories. He allows this bias to color his work. I think he'd be better off dropping the negative tactic and adopt a positive approach. He could focus on what is wonderful about the scientific account of the creation story and avoid denigrating the other versions.

His entire treatise may be undermined by his attitude of damning the opposition, which is the way of politicians. Such a mentality turns people off of religion and could turn people away from Hawking's cosmology.

The answer to question three "Is there only one set of possible laws?" turns out to be the thesis of Hawking's book. He believes there are infinite sets of possible laws. He will deal with this in a later chapter, he says, so we'll come back to it then.

The next idea offered in *The Grand Design* is that there's no free will. Nobody has the freedom to choose what they do or say.

> "Biology shows that biological processes are governed by the laws of physics and chemistry and therefore are as determined as the orbits of the planets. Recent experiments in neuroscience support the view that it is our physical brain, following the known laws of science, that determines our actions, and not some agency that exists outside those laws."

Apparently, we don't control our brain because our brain controls us. Our choice to visit Paris or Rome or China is determined by known laws of science. Hawking never reveals which known laws of science determine what I eat, what I wear, and who I like or dislike. Thus, his premise remains in the realm of conjecture until he supplies the relevant data.

Professor Hawking states categorically that he's a scientific determinist. He believes the laws of science determine everything that happens in the past, present, and future. Again, this mentality mimics fundamentalist theists who say that God determines everything past, present, and future.

By denying free will and saying we are forced to act according to natural laws, Hawking affirms the existence of a higher power that has control over human beings. The "higher power" in this case is the laws of physics, so this must be his "religion."

He applies this determinist hypothesis to human nature and concludes there's no free will because everything is determined by scientific laws. But the question of the origin of the laws is conveniently ignored. Again, we are in the realm of conjecture, not scientific knowledge.

He never reveals which laws he is talking about, yet he declares: "It is hard to imagine how free will can operate if our behavior is determined by physical law, so it seems that we are no more than biological machines and that free will is just an illusion."

A major problem with his determinism idea is this: we have to accept that there's no such thing as a criminal. People who choose to commit violent crimes are merely biological robots who are forced to act by the laws of physics. This would include Hitler, Stalin, serial killers, and suicide bombers.

I wonder if Hawking can live his philosophy. Would he call the police if burglars came into his home, or if a gang tried to molest his daughter? Or, would he acknowledge they are only acting under scientific law?

Hawking's theory has zero value if the consequences don't apply to real world situations. Then his theory is just a mental construct, nothing else.

Moreover, if laws of physics determine everything, then his own free will in writing his book is called into question. He can no longer be confident that what he writes is true, because it's coming from some unspecified law of physics acting upon him.

To live by his understanding, he should delete his name from his book and put "written by an unspecified law of physics."

Ironically, the same physical law acts upon a theist who writes that God created the universe. The same law presents contradictory conclusions? Clearly this is nonsense.

What does make sense is that every person speaks according to his own realization. In his own words Hawking says *he* cannot imagine "how free will can operate if our behavior is determined by physical law." This is the understanding of Stephen Hawking.

Of course, that doesn't exclude *others* from understanding how free will can operate despite physical law. Herein lies the fault of every skeptic. His own mindset is his binding truth. Whatever he can understand is his entire understanding—in spite of what others understand.

Hawking needs to provide some evidence for his conjecture. Otherwise his opinion is "blind faith in the absence of evidence" according to Richard Dawkins.

I challenge Hawking's argument with a thought experiment a la Einstein.

My body follows the laws of nature in all its functioning. In terms of chemical reactions, there's no law of nature it is ignoring. Yet, I have a will and intelligence over my body. If I go for a swim, all the chemical reactions that govern breathing, heartbeat, and the flow of oxygen in my body change accordingly. These will all be different again if I'm reading a book in a library.

By the exercise of my free will I can manipulate my atoms and change the functioning of my bodily machinery according to a specific purpose I want to accomplish. And yet, no physical law is ever broken or even modified.

Clearly, my mind does have a will over my body which functions according to the laws of nature. When we consider the accomplishment of human life which is intelligent enough to examine its origins, it would be ludicrous to think that I have no free will.

Hawking should give more thought to this matter. He needs to answer these questions: Which physical law determined Mozart's or Beethoven's music? Or the music of Elvis Presley, or the Beatles?

Which physical law determined the religious fervor of Mother Teresa, or St. Francis of Assisi? Does the same physical law convince some people that God exists, and other people that God doesn't exist? If so, this physical law can't make up its mind.

Some readers will guess that Hawking is referring to ideas from biology, like adaptation/conditioning. They may buy that the laws of physics govern free will because they determine the laws of chemistry, which determine the laws of biology, which determine the laws of psychology, which determine the laws of sociology. They might argue that opinions about beauty and genius are simply conventions from adaptation and conditioning which are "explained" by biology, which relies on biochemistry, which relies on physics.

Mother Teresa's compassion might be seen as a favorable disposition to keep society from driving itself to extinction. People who believe in God would be dismissed as simply naïve, or maybe less-evolved people who will eventually die out along with their creation stories. Dishonest people are biologically, genetically, or circumstantially driven to such behavior. Ultimately, the laws of physics prevail.

However, people who posit such arguments still insist that criminals who molest their children be dealt justice. Their theories and opinions are creations of their own views, which in practice may not be useful.

The problem, of course, is that no scientist could ever provide data to substantiate the non-existence of free will, due to the insurmountable complexity involved in determining action. So it remains in the jurisdiction of speculation.

When we discuss science there is a need to substantiate hypotheses with data. Without evidence we have conjecture, or its close cousin blind faith, both of which live in the "no data" region of existence.

If Hawking is certain that some physical law determines whether people become honest rather than dishonest, or vice versa, and that they are driven to such behavior, then he needs to specify which physical law he has pinpointed to relieve our convicted felons of their guilty status.

Unfortunately for the convicts, his *Grand Design* does not specify any such law. Hawking is losing credibility with me. He has a particular mindset that he projects onto the picture of reality, and thereby constructs a "theory" to substantiate what he *wants* to be true. But with no data we're left with blind faith, and that's not science. Maybe it's reverse engineering in reverse.

I'm curious to know the answers to his three "why" questions, so our lawyer agrees to plod on.

Third Chapter

At this point, Hawking is going to explain the true meaning of objective reality. What we think is reality, he says, may only be the simulated reality of websites such as *Second Life*.

He admits that his idea comes from the science fiction film *The Matrix*, "in which the human race is unknowingly living in a simulated virtual reality created by intelligent computers to keep them pacified and content while the computers suck their bio-electrical energy."

His book offers various concepts and models of reality, but most are science fantasy. None of his ideas are new. They can be found in science fiction, Zen Buddhism, and yoga philosophy. Most yoga adherents know what is Illusion, *maya*, and Truth, *tattva*.

This chapter is just the same old wine in a different bottle.

There is no acknowledgment that ideas have already been dealt with by other traditions. Moreover, these ideas are philosophical, not scientific.

To the informed reader, the modern jargon unsuccessfully disguises these ideas as fresh. Where are the new ideas? We don't want a rehash or another *Matrix* sequel.

Yoga philosophy offers a positive contribution to the realm of thought. We may live in an illusory objective world where everything is temporary, but a true reality does indeed exist.

The very concept of illusion relies on the fact of an actual reality. For example, mistaking a piece of rope for a snake implies that snakes exist. The goal of life, according to yoga philosophy, is to realize true reality through the yoga process. That's how we escape illusion.

When knowledge appears, ignorance disappears. When the Sun appears, darkness vanishes. A rope can only be mistaken for a snake in the dark. True reality is a different nature than an experience that props up the illusion.

In contrast to yoga philosophy, Hawking states that we can't know reality. Does this mean we must remain in illusion forever? Again, this is the rationale of the skeptic. If he can't know reality, neither can anyone else.

In spite of this, he will now provide a model to help us understand the unknowable. First, he explains the four qualities of a good model:

1. A good model is elegant

2. It agrees with and explains all observations

3. It has few arbitrary or adjustable elements

4. It can make predictions about future observations that disprove the model if they are falsified.

Then he offers a perspective which is important to understand his book:

> "There is no picture- or theory-independent concept of reality. Instead we will adopt a view that we will call model-dependent realism: the idea that a physical theory or world picture is a model (generally of a mathematical nature) and a set of rules that connect the elements of the model to observations."

A particular understanding of reality is based on some particular model, he says. But reality is independent of any model thus far designed by physicists. They are constantly adjusting their models.

Hawking assures us that model-dependent realism "can provide a framework to discuss questions such as: If the world was created a finite time ago, what happened before that?" Interesting. But does it relate to the real world in which we all live?

Apparently not. "According to model-dependent realism it is pointless to ask whether a model is real," he says, "only whether it agrees with observation."

Yet, earlier in his book, Hawking discussed the model of Ptolemy and dismissed it even though it *did* agree with observation. Ptolemy's model was rejected even though Hawking assured us that "it is pointless to ask whether a model is real, only whether it agrees with observation." Evidently, the old professor doesn't follow his own advice.

However, he does admit that it is not "clear yet whether a model in which time continued back beyond the Big Bang would be better at explaining present observations because it seems the laws of the evolution of the universe may break down at the Big Bang."

Although he posits a framework to discuss what happened before creation, he concludes that even if we had such a model it might not help us. Why? Because the laws that govern the evolution of the universe evidently break down at the point of creation. He's not certain, but Einstein's equations "may break down at the Big Bang." That's his informed opinion.

Let's take a second look at Hawking's PhD thesis. Perhaps you remember that he said, "I was awarded my PhD by showing that Einstein's theory of relativity implied that the universe must have begun with a Big Bang."

By going backwards in time, he posited that the universe would resemble the collapsing star of Roger Penrose's theory. Penrose developed the singularity theorems that Hawking borrowed to prove that by going back to the Big Bang we arrive at a singularity.

But now in the 21st century, going back in time means physical laws break down at the Big Bang. Mimicking Penrose's theory is no longer a viable concept to understand the "bang" in the Big Bang.

We are left with this issue: if physical laws break down as we approach the Big Bang, how do we know they even existed prior to that? We accept that cosmic inflation defied all known laws of physics. Then, what caused inflation?

If inflation was not caused by any known laws of physics, does it imply that the laws at creation were supernatural in origin? Or, did the laws simply arise after creation and not before? If so, what is that mechanism that manufactures universal laws?

If we were hoping for clarity in regards to reality, it looks like another speculative walk down the garden path. But Hawking will stick to his guns because "physicists are indeed tenacious in their attempts to rescue theories they admire..."

Is "rescuing theories they admire" the job description of physicists? The word "rescue" means saving someone or something from peril. The imperiled theory Hawking intends to rescue appears to be the modern creation account.

He cites former models of the universe that fell by the wayside, "such as the four-element theory, the Ptolemaic model, the phlogiston theory, the Big Bang theory, and so on. With each theory or model, our concepts of reality and of the fundamental constituents of the universe have changed."

In due course, the static universe model was rejected, and the Big Bang theory became the accepted model by the mid-twentieth century. We know that the physics of Newton gave way to Einstein's physics.

Clearly, by including the Big Bang on the list of redundant theories, Hawking admits that it is no longer viable. He assumes that his Grand Design model will render the Big Bang model obsolete. No lack of bravado on his part. However, it means his PhD thesis was a conjecture that looked true at the time, but no longer looks true. It's no longer a substantial contribution to knowledge. Should he renounce his PhD title now?

We may note that reality remains the same, as it was before and as it is now. Only our ability to observe it has changed, and thus our conception. Next, he claims the following:

> "In the 1920s, most physicists believed that the universe was static. Then, in 1929, Edwin Hubble published his observations showing that the universe is expanding."

Did Hubble really believe the universe was expanding? I deal with this extensively in Chapter Twelve.

Regarding the laws that govern the universe, Hawking concludes: "There seems to be no single mathematical model or theory that can describe every aspect of the universe. Instead, as mentioned in the opening chapter, there seems to be the network of theories called M-theory. Each theory in the M-theory network is good at describing phenomena within a certain range."

So it seems the M-theory model will challenge the Big Bang model. Hawking believes that Ed Witten's M-theory is the answer, although Witten's name is never mentioned. Instead, the model is called the Grand Design.

Hawking is unaware we have researched M-theory and discovered it's merely a set of ideas, hopes, and aspirations. It can't be called a network of theories, as defined by genuine scientific methodology. If we accept the statements of reputed physicists themselves, it's more like a herd of hopeful hypotheses.

Ironically, even a network of theories is inelegant. It's similar to Ptolemy's epicycles. This defeats the first premise of a good model. Of course, if we accept his framework he can lead us to his own conclusions. But if his framework is faulty, or if new discoveries demonstrate his model to be inaccurate, then we end up with ignorance, not knowledge.

Hawking is full of hope that the grand design model will not fall by the wayside. At this point, however, I'm not as hopeful.

For the rest of his book Hawking uses his own brand of philosophy, not science, to support model-dependent realism. Again, reality pre-exists any model he can imagine. He is trying to shoehorn reality to fit his own ideas about M-theory.

This mentality of gambling with "knowledge" which can lapse into ignorance within a few decades is of little benefit to humanity.

Hawking's logic in this chapter is not sound, and that doesn't bode well for the rest of his book.

6
A Theory of Everything?

So far in *The Grand Design* we have encountered contradictions, along with speculations that lack evidence. Is there more of the same?

Our attorney questions whether we will finally get some genuine physics, and perhaps observational data to substantiate the conjecture.

Fourth Chapter

The next chapter begins by stating that Newtonian physics was inadequate to describe how nature behaves at the atomic and subatomic level. Thus, the newer model of quantum mechanics. For Hawking, the fact that atomic particles behave differently than described by Newton's laws means we can no longer rely on intuitive logic.

His justification is that reality "is a picture in which many concepts fundamental to our intuitive understanding of reality no longer have meaning."

He remarks that the components of normal objects, atoms and molecules, obey quantum physics. But composite objects like tables and chairs behave according to Newton's laws. Newton's laws match the view of reality we experience in everyday life. Atoms and molecules reveal a different version of reality. So again, we are asked to abandon common sense.

By definition, common sense does not apply to objects which are beyond our purview, like atoms. It is only useful for the everyday version of reality. So Hawking's idea to abandon common sense for a different version of reality to which it does not even apply is not coherent.

The traditional creation story also defined God as a different version of reality, but the concept of a supernatural entity applied. This is not what Hawking is talking about.

"Physicists are still working to figure out the details of how Newton's laws emerge from the quantum domain," he admits. "What we do know is that the components of all objects obey the laws of quantum physics, and the Newtonian laws are a good approximation for describing the way macroscopic objects made of those quantum components behave."

At this point several features of quantum physics need to be explained:

1. The wave/particle duality. "That matter particles behaved like a wave surprised everyone."

2. The Heisenberg Uncertainty Principle, formulated in 1926 by Werner Heisenberg, explains that our ability to simultaneously measure the velocity and the position of a particle is severely limited.

3. The Uncertainty Principle states: "if you multiply the uncertainty in the position of a particle by the uncertainty in its momentum (its mass times its velocity) the result can never be smaller than a certain fixed quantity called Planck's constant."

4. The more precisely we measure the position of a particle the less accurate is its velocity, and vice versa.

If we double the uncertainty for speed, then we will need to halve the uncertainty for position. This means that, "given the initial state of a system, nature determines its future state through a process that is fundamentally uncertain."

The above explanations might not be readily comprehensible, but you can feel secure in the knowledge that quantum theory also bothered Einstein. That's why he became critical of it.

Richard Feynman, another great scientist of the mid-20th century, wrote, "I think I can safely say that nobody understands quantum mechanics." Are physicists any closer to that understanding today?

Scientific theory has two components: the mathematical core, and the physical interpretation of the mathematical equations. Quantum mathematics is substantiated, but the physical interpretation is so obscure that Einstein and Feynman had to speak up.

Hawking and Mlodinow, however, assure us that they understand quantum mechanics. Quantum physics might seem to sabotage the laws of nature, but Hawking affirms that nature determines the *probabilities* of sub-atomic events rather than choosing a definite outcome. He then extends the indeterminate action of sub-atomic particles to the universal scale and makes a claim for a new universal determinism.

This idea "leads us to accept a new form of determinism," he writes. "Given the state of a system at some time, the laws of nature determine the *probabilities* of various futures and pasts rather than determining the past and future with certainty."

Of course, quantum physics only deals with the sub-atomic level. On the level of reality of everyday life, there is the thoroughly predictable Newtonian physics based on laws that govern familiar matter. Hawking has already admitted the difference but now passes the uncertainty of the sub-atomic level onto the scale to which it does not belong—familiar matter which we can perceive. Doesn't he see the contradiction?

Apparently not, because he asserts that quantum theory should be interpreted in such a way that the past is no longer fixed history as we generally believe. Hawking's book acknowledges that some people may find this idea offensive. Others may find it ridiculous.

Can this interpretation be verified? Of course, it has never been tested. Again, the reader is asked to abandon sound judgment. Now, he claims, "Scientists must accept theories that agree with experiment, not their own preconceived notions."

Good point. Experiment must verify theory. Can Stephen Hawking live up to his claim? Has any macro-system of matter, from a marble to a galaxy, ever been proven experimentally to behave as sub-atomic particles do? Hawking does not provide any evidence.

Moreover, the Uncertainty Principle of quantum physics presents a major interpretive conflict. It proposes that the movements of particles in sub-atomic reality behave in a manner that is uncertain, but that uncertainty could just as well be generated by our lack of understanding.

If we can never accurately measure both velocity and position of a particle simultaneously, it might mean that the movements of the particles are indeterminate. It could also mean that the observer was unable to determine their behavior due to a lack of knowledge.

Einstein and other physicists postulated the "hidden variable theory" to recognize incomplete knowledge. They implied that the "Copenhagen interpretation" of quantum theory by Neils Bohr was an incomplete description of reality. They felt there was a more fundamental theory hidden beneath quantum mechanics whereby the entire system would reveal a deeper reality which could predict outcomes with certainty.

Quantum physicists were quite certain about the nature of uncertainty, so the hidden variable idea lost popularity over time. But now theorists are exploring ways to bring back the idea with modifications because it's foolish to assume that we know it all and there is nothing left to be discovered.

To reduce the burden of uncertainty on sub-atomic particles, Hawking assures us that if an experiment is repeated enough times the data will represent the probability of a particle's position and velocity. This would make it scientifically valid. He accepts Richard Feynman's theoretical sum-over-histories model, and uses it as a practical paradigm to try and solve the indeterminacy problem.

According to Feynman, a system takes all possible paths which sum up to a probability amplitude. But there is no determinism in the sense that there is no specific answer at the end, just probabilities. The result is we are not left with a certain event, only a theoretical probability—an approximation of outcomes—because no one can know all the possibilities. Nor can one even assume that all the possibilities are observable. That's why the word "probability" is appropriate and not the word "certainty."

Generally, books and papers on quantum theory discuss the various interpretations and observable effects, but also state that science does not know the correct answer at present. Stephen Hawking misrepresents the *status quo* of quantum physics by presenting that science has moved forward in its understanding of quantum mechanics and an established accepted understanding does exist.

Basing a grand scientific conclusion on a particular explanation without acknowledging there are deficiencies in the argument, and there are other valid alternative explanations, misleads people to accept that one particular view is actually correct.

Furthermore, interpretation is not the same as the knowledge or proof of a thing. Cosmologist Sean Carroll describes the problem of interpretation in quantum mechanics as the "greatest embarrassment" to modern physics. He says he's embarrassed because physics is looking less like science and more like philosophy (where everybody gets to put in their two cents worth).

Until there is data to distinguish between interpretations, each version is like a beauty pageant contestant striving to emerge victorious. Needless to say, that is not science.

Theorists, however, can take advantage of this new face of science where interpretation comes forward and proof is left on the back burner. It means they can publish hundreds of papers and nobody can prove them wrong.

Quantum mechanics is a wonderful theory. It can explain things that were unexplainable previously. However, Roger Penrose points out in the September 2009 issue of *Discover* magazine, "when you accept the weirdness of quantum mechanics [in the macro world] you have to give up the idea of space-time as we know it from Einstein. The greatest weirdness here is that it doesn't make sense. If you follow the rules, you come up with something that just isn't right."

The quantum description of reality appears contrary to the reality we experience. Quantum theory postulates that a single object can be in many places simultaneously. Penrose takes issue with this idea in his *Discover* interview:

> "It doesn't make any sense, and there is a simple reason. You see, the mathematics of quantum mechanics has two parts to it. One is the evolution of a quantum system, which is described extremely precisely and accurately by the Schrödinger equation. That equation tells you this: If you know what the state of the system is now, you can calculate what it will be doing 10 minutes from now."

> "However, there is the second part of quantum mechanics—the thing that happens when you

> want to make a measurement. Instead of getting a single answer, you use the equation to work out the probabilities of certain outcomes. The results don't say, 'This is what the world is doing.' Instead, they just describe the probability of its doing any one thing. The equation should describe the world in a completely deterministic way, but it doesn't."

Erwin Schrödinger is regarded as a genius; accordingly did he recognize the conflict his equation created? Penrose has no doubt about this:

> "Schrödinger was as aware of this as anybody. He talks about his hypothetical cat and says, more or less, 'Okay, if you believe what my equation says, you must believe that this cat is dead and alive at the same time.' He says, 'That's obviously nonsense, because it's not like that. Therefore, my equation can't be right for a cat. So there must be some other factor involved.'"

Clearly, if it's not right for the cat, then it can't be right for other macro objects. Penrose confirms that this is what Schrödinger is pointing out.

To accept that the cat is dead and alive at the same time is a reality that no one has ever experienced. Actually, it's a superimposition of two contradictory conditions existing simultaneously. Yet theoretical physicists justify this bizarre cat analogy as a reality we must acknowledge to rationalize that our consciousness can take other paths without our knowing it. This begs the question, how can we verify what can't be experienced or even known?

Yet this conception is driving many ideas in theoretical physics towards the many worlds interpretation of reality. This interpretation posits that all probabilities exist somewhere in some parallel universe. Penrose says, "You're led to a completely crazy point of view. You're led into this many worlds stuff, which has no relationship to what we actually perceive."

Stephen Hawking should honestly admit that his theory is really just an interpretation until it's verified by data. He also fails to mention the mathematical problems involved in his interpretation. By not being completely candid, he leads his readers into accepting what Einstein and Feynman did not accept.

Consciousness

Richard Feynman wrote that the famed double-slit experiment "contains all the mystery of quantum mechanics."[16]

[youtube.com/watch?v=DfPeprQ7oGc]

When particles pass through two slits, they form a pattern of two lines on a back screen accordingly. When waves pass through the same two slits, they form a series of lines of different intensities on the back screen called an interference, or diffraction, pattern. However, when electrons pass through the double slit, they imitate the trajectory of waves.

How could particles of matter create an interference pattern like waves? Physicists decided to place a measuring device on one slit to detect what happens when each electron passes through it. Then the mysterious nature of the subatomic world was revealed. Suddenly, the electrons acted as particles and formed two neat lines. Did they recognize they were being observed?

Feynman came to the amazing conclusion that the particles in the double-slit experiment had some sort of information. As soon as an observer tried to track the particles, they behaved differently. The act of measuring collapsed the wave function.

Light had always been accepted as a wave, but Einstein said

---

[16] www.youtube.com/watch?v=DfPeprQ7oGc

that light was a particle, a photon, and his theory supported that conclusion. Yet when light goes through the double slit it forms a diffraction pattern of many bands, like a wave does, so how could it be a particle?

Once again, physicists set up a measuring device to find out what's going on. They fired photons one at a time through the double slit and this time the result was two lines, as would be expected from a particle. Again, the act of measuring collapsed the wave function. Were the detectors interfering?

Heisenburg explained that the experimental apparatus itself alters the sub-atomic particles' trajectory because the particles we are trying to observe are of the same scale as the photons we're using to observe it. In other words, to observe something we must bounce photons off it. But the photons disturb the particles because they are the same size. So there is no way to observe sub-atomic particles without altering their trajectories.

The double-slit experiment was repeated with the detectors still in place but with no data being recorded. The detectors were detecting but not collecting the data. This time, the photons made a diffraction pattern.

To recap: when physicists collect data, photons act like a particle. If they don't collect data, photons act like a wave. Does the conscious act of collecting data determine whether light behaves like a wave or a particle?

Erwin Schrödinger posited that since the photons don't break up, they must simply exist as a probability distribution, rather than a particle, before any measurement is made. Therefore, as waves of probability they interfere with themselves as they pass through the slits and cause a diffraction pattern on a back screen. When no measurement is made, there is just probability. However, when a detector is used at the slits, the probability wave function collapses to a physical particle.

Clearly, in the double-slit experiment there seems to be two laws at work: one that applies when you're detecting, and another that applies when you're not detecting. This attaches some importance to the awareness of the observed as well as the observer. In other words, reality seems to be a product of some form of consciousness, as noted by Nobel Laureate Eugene Wigner.

> "It will remain remarkable, in whatever way our future concepts may develop, that the very study of the external world led to the scientific conclusion that the content of the consciousness is the ultimate universal reality."

Max Planck echoed the same realization as Wigner, "Science cannot solve the ultimate mystery of nature because, in the last analysis, we ourselves are a part of the mystery that we are trying to solve."

Even more amazing is the mysterious phenomenon of entanglement of particles.

[youtube.com/watch?v=1BfJ06plOTs]

When two atoms are entangled with one another they share a peculiar connection that is instantaneous and unaffected by distance. Experiments establish that even over a vast stellar expanse, the atoms remain connected. Whatever happens to one instantaneously affects the other. This behavior is so synchronized, like a choreographed dance, that it stretches the concept of communication. They are so connected, that the word "entanglement" is used. Einstein disliked the concept, calling it "spooky action at a distance."

Of course, if particles can transfer information we must ask, how do they acquire it, how do they transmit it, and where does the information come from? Such knowledge implies some form of intelligence, which means some form of consciousness.

Schrödinger affirmed that entanglement, which is the concept of connectivity, is factually the basic property of quantum mechanics. Quantum mechanics, then, is the display of waves of potentiality. It's the field of pure potentiality, of abstract potential existence. Connectivity among all things is a basic constituent of the fabric of reality.

Feynman's interpretation avoided the conclusion reached by Wigner and Planck. Sidestepping the taboo subject of consciousness, he introduced a different perspective: that the particles take every possible path rather than taking one definite path. Moreover, they seem to take every possible path simultaneously. But this offered no explanation why particles behave differently when measured.

It's clear that humans can't know every possible outcome, yet it appears that particles can, according to Feynman. Thus, we again come to Wigner's conclusion of reality being a product of some form of consciousness. Feynman undermines Hawking's point that the universe simply popped out of nothing if it contains all the symptoms of intelligence, including particles with prior knowledge and finely-tuned laws of mathematics and physics.

Alluding to Feynman's explanation, Professor Hawking writes: "There are an infinite number of paths, which makes the mathematics a bit complicated, but it works."

The Grand Design theory must be his personal conviction that it works, but the concept of infinity does more than make the math just "a bit complicated". Every mathematician knows, and Hawking also knows, that we cannot do mathematics with infinity.

A true infinite has no beginning and no ending. Mathematicians may say there are infinite numbers in the set from 2 to 8. But that infinite has bounds, a beginning 2 and an end 8. Moreover, the numbers 1 and 9 are not included in this infinite set. But there is no question of doing mathematics with an actual infinite.

Physics studies physical things, which have a beginning and an end. The Big Bang concept establishes that the universe is physical because it had a beginning. And the particles of quantum physics had their beginning also. An actual infinite, on the other hand, means something beyond physical nature, a supernature without beginning or end.

By the following four tenets physics and cosmology paint themselves into a corner in which the modern creation story seems trapped:

1. Physical nature is all that exists
2. Supernatural events do not exist
3. M-theory posits infinite natures
4. These natures can never be supernatural

Modern science doesn't allow anything supernatural to enter its story because that would grant some validity to the traditional story. Yet in contrast to premises 2 and 4, the existence of dimensions beyond our present understanding is a possibility according to M-theory. There's no data to prove it now but like dark energy, it could well be discovered.

Fifth Chapter

The title of Hawking's fifth chapter is "The Theory of Everything." Really? Not just physics stuff, but everything else like cures for disease, old age, and death? Surely these are included in everything.

The chapter begins with an Einstein quote: "The most incomprehensible thing about the universe is that it is comprehensible."

Professor Hawking concludes this is because the universe is governed by "scientific laws." With all due respect, the old professor is not expressing himself properly. The entire cosmos is gov-

107

erned by laws of nature. Hawking's presentation is misleading.

For example, almost a dozen theories of electromagnetism existed up to the 1860s and all of them were flawed. Then, James Maxwell discovered the mathematics that explained the relationship between light, electricity, and magnetism. Hawking writes, "Maxwell had unified electricity and magnetism into one unified force."

Factually, Maxwell did no such thing. He was able to demonstrate that it was *already* one unified force. Maxwell's "laws" already existed and governed universal forces billions of years before Maxwell was born.

The fact that various branches of science are discovering natural laws hardly makes them the laws of science. Columbus may have discovered America but no one will claim it's the land of Columbus. Like America, nature's laws existed for millennia before science or Columbus claimed credit. Moreover, if our understanding of the laws changes over time, nature does not oblige us with different behavior accordingly. Once again, the old professor isn't expressing himself properly.

Nevertheless, in 1687 Isaac Newton presented the law of gravity as a mathematical formulation: Every object in the universe attracts every other object with a force proportional to its mass; the familiar Inverse Square law of gravity.

Subsequently, scientists discovered that various phenomena of the universe could be modeled if the mathematical machinery was understood. They were discovering mathematical formulations like explorers discovering new lands, and claiming proprietorship over the same.

The question remains: How did the laws of physics and mathematics first come into being so that scientists could discover their existence?

Unfortunately, Hawking skirts the issue of the origin of natural laws. He sweeps it under the rug and pretends that the house of physics is clean. But he needs to explain why these universal laws exist, because he has promised this from his first chapter.

At this juncture in his book, Hawking introduces string theory.

> "According to string theory, particles are not points, but patterns of vibration that have length but no height or width—like infinitely thin pieces of string. String theories also lead to infinities, but it is believed that in the right version they will all cancel out. They have another unusual feature: They are consistent only if space-time has ten dimensions, instead of the usual four."

The strings are one-dimensional, but the math only works if there are ten dimensions? Of course we only observe three plus time, which makes four dimensions—Einstein's space-time.

Physicists are leading us from the unobservable to the inconceivable, and now to the unprovable. If that doesn't complicate our picture of "reality" enough, we find the mathematics points to not one, but five possible string theories.

"String theorists are now convinced that the five different string theories and super-gravity are just different approximations to a more fundamental theory, each valid in different situations," explains Professor Hawking.

The more fundamental theory he refers to is M-theory, which was mentioned earlier: "People are still trying to decipher the nature of M-theory, but that may not be possible."

It may not be possible to decipher M-theory, but Hawking will give it a try. This is the first disclaimer for his idea of a grand unified theory of everything using M-theory.

Another disclaimer: "It could be that the physicist's traditional expectation of a single theory of nature is untenable, and there exists no single formulation."

The third disclaimer: "It might be that to describe the universe, we have to employ different theories in different situations."

How then is it one unified theory as claimed? It seems Hawking's assertions follow the quantum laws of uncertainty. We can name these disclaimers: the Hawking Uncertainty Principle of M-theory.

> "Each theory may have its own version of reality, but according to model-dependent realism, that is acceptable so long as the theories agree in their predictions, that is, whenever they can both be applied."

One grand unified theory (GUT) may not be possible. Nonetheless, Hawking suggests utilizing different theories for different aspects of the universe and still have a unified theory. Of course, it won't be as elegant as one GUT. In fact, far less elegant than Ptolemy's model which was rejected even though it gave accurate approximations of orbital motion.

Hawking is sold on M-theory and promotes it vigorously. As he stated earlier, "physicists are indeed tenacious in their attempts to rescue theories they admire…"

In his book Hawking acts like a lawyer who represents M-theory just like he did on the "Larry King Show." Apparently, it needs defending because reputable physicists affirm that it is untestable, thus stretching the boundaries of good science.

Still, Hawking would have us believe that the theory has now become law. In his own words: "The laws of M-theory therefore allow for *different universes* with different apparent laws, depending on how the internal space is curled. M-theory has solutions that allow for many different internal spaces, perhaps as many as 10^{500}, which means it allows for 10^{500} different universes with its own laws." [his emphasis]

We have already ascertained that M-theory is simply a collection of ideas, yet Hawking presents these ideas as "the laws of M-theory." But if the goal is to unify all natural laws into one GUT, M-theory is a giant leap in the wrong direction when it predicts 10^{500} different possible worlds, all with different laws, only one of which would be our universe.

Yet Hawking's book makes an appeal for this multiverse concept, a conception which multiplies the probability that the fine-tuning of the cosmos came about by chance. Let's examine the concept.

We know the early conditions after the Big Bang explosion were finely calibrated for intelligent life to arise. If conditions are adjusted by the tiniest amount, life would be impossible. Hawking doesn't deny fine-tuning. He even qualifies it as almost miraculous. The universe does look like it's been exquisitely designed to produce intelligent life. Hawking won't dispute that. But after admitting it looks like that, he explains it via the multiverse idea.

Does the unlimited universes idea account for fine-tuning? With unlimited roulette wheels the chance that my number will come up on at least one of them is greatly enhanced. With unlimited universes, each having a unique set of laws, the chance that one universe is fine-tuned for life, or bio-friendly, is a reasonable expectation from the probability laws. And that universe might just as well be ours.

However, this is simply a trivial solution, not a rigorous proof, because we cannot prove the existence of these other worlds, what to speak of examining their laws to see if they are bio-friendly. Such conjecture, however, has a respectable name: it's called the weak anthropic principle.

The multiverse conception postulates a set of universes rather than one universe (uni=one, multi=many). It came about by physicists speculating that instead of our one universe there might be various different cosmic regions resembling a patchwork quilt of universes with differing physical laws and properties.

From this hypothesis, Leonard Susskind of Stanford University tried to resolve the Anthropic Principle, the situation of our universe being fine-tuned for life, in his book *The Cosmic Landscape*. He posited that life obviously couldn't survive in a cosmic region where life was impossible, so it's not astonishing that life thrives in our corner of the multiverse. Thus, he created the weak anthropic principle.

Susskind's conjecture was happily greeted by scientists who weren't happy that the fine-tuning of the universe evoked the dreadful God concept. Paul Davies describes it like this; "...they seized on the multiverse theory as a neat explanation for the uncanny bio-friendliness of the universe."

Apparently, Hawking anticipated an outrage for his use of the multiverse to explain fine-tuning. Thus, we find another disclaimer in his seventh chapter. The "multiverse idea is not a notion invented to account for the miracle of fine-tuning. It is a consequence of the no-boundary condition as well as many other theories of modern cosmology."

The "no-boundary condition" refers to the assumption that the universe is infinite which would allow for a possible multiverse to exist. But then we find an inconsistency in Hawking's same chapter. Although he claims the multiverse idea was not invented to account for the miracle of fine-tuning, he does use it to account for the miracle of fine-tuning.

He writes: "in the same way that the environmental coincidences of our solar system were rendered unremarkable by the realization that billions of such systems exist, the fine-tunings in the laws of nature can be explained by the existence of multiple universes."

There is no data to support "the existence of multiple universes." It's an unproven hypothesis that Dr. Hawking arbitrarily elevates to genuine existence. Does an unproven hypothesis validate multiple universes? Clearly not. Yet Hawking does prop up a so-called "realization" to promote his grand design model in the next example.

> "Just as Darwin and Wallace explained how the apparently miraculous design of living forms could appear without intervention by a supreme being, the multiverse concept can explain the fine-tuning of physical law without the need for a benevolent creator who made the universe for our benefit."

So, the multiverse concept *is* promoted as a notion to account for fine-tuning. Hawking's later statements contradict his earlier ones. It's one contradiction after another. This sleight-of-hand makes his theory look like a swindle.

So now we will examine Hawking's claims in more detail. Does the multiverse concept factually explain the fine-tuning of the universe? And since he brought it up, does it rule out a benevolent creator in the process?

If M-theory predicts unlimited possibilities it means anything and everything is possible. Therefore, if a theologian claims that God created a universe where Jesus Christ died for people's sins, science can no longer deny this possibility. M-theory allows for it in a scenario with infinite universes and infinite possibilities. Hawking's model can't escape the logic that the greatest likelihood of this event was in our own universe.

Some may argue that the probabilities M-theory proposes refer to different laws of nature as well as different parameters in the laws (different particle masses, charges, etc.). It does not predict universes where a supernatural force may be present. But this argument, that M-theory can predict whether or not a force is supernatural, is an arbitrary assumption not substantiated by experimentation, observation, or any method accepted as scientific.

Essentially, the multiverse concept takes aim at the teleological use of the Anthropic Principle, which suggests that the cosmos is fine-tuned to produce intelligent life. Hawking is trapped. He can't banish the possibility of teleology and still argue that infinite possibilities exist.

The Anthropic Principle does indicate teleological laws at work in our observable universe. Cosmologists have to deal with this issue if infinite possibilities exist.

Basically, the multiverse idea is an untested assumption so it lacks scientific merit. It simply adds unlimited roulette wheels so my number will come up. When speculative ideas are needed to make theories palatable, we have the right to question the credibility of such theories.

Roger Penrose has raised significant criticisms of the many worlds hypothesis to account for fine-tuning. Yet Hawking doesn't respond to any of these published critiques. To completely ignore the criticism of a physicist of Penrose's stature could be seen as deliberate blindness.

Invoking imaginary worlds and unsupported philosophical claims without doing the hard work of backing up the talk with valid data is hardly science. It reminds me of the unsupported realms of heaven and hell. In this regard, there's no difference between religion and science to an astute observer.

So far, *The Grand Design* has presented us with speculative ideas. Does M-theory with its unlimited universes represent the real world?

Taken at face value, Stephen Hawking's hypothesis seems to be half-baked. Any lawyer could reasonably argue that without evidence it's too far-fetched and deserves to be exposed as unverifiable.

7
Choose Your Universe

The title of Hawking's next chapter is "Choosing our Universe." (Yes, Virginia, that is the title). Obviously, the universe was already chosen for us before any human was born. So where is Stephen Hawking going with this?

We already know that Professor Hawking doesn't attach any importance to whether a model represents reality. In his book he stated, "According to model-dependent realism it is pointless to ask whether a model is real, only whether it agrees with observation."

So the validity of his M-theory model hinges entirely on whether it agrees with observation. But is it possible to observe these limitless universes, all with different laws, and even with ten dimensions? We can only observe a small part of our three-dimensional universe. This is the argument our lawyer will present to the jury.

SIXTH CHAPTER

Dr. Hawking's Sixth Chapter comments on the Boshongo creation myth and the god Bumba, before looking at the Mayan creation story. It's easy to use the creation myths of unfamiliar African tribes and extinct cultures to contrast the modern creation story and enhance its appeal.

Of course, he doesn't use the Genesis story as an example—too many people might be offended. Nor does he use the Koran as an example—way too dangerous. But it's clear that he views the Bible and Koran creation stories as much a superstitious myth as any African origins story.

"Creation myths like these," he explains, "attempt to answer the question we address in this book: Why is there a universe, and why is the universe the way it is? Our ability to address such questions has grown steadily in the centuries since the ancient Greeks, most profoundly over the past century. Armed with the background of the previous chapters, we are now ready to offer a possible answer to these questions."

At last. Well, better late than never. Yet for some unknown reason his book goes off on a tangent. Instead of offering the possible answer we've been promised, Hawking distracts his readers with the following argument.

> "One thing that may have been apparent even in early times was that either the universe was a very recent creation or else human beings have existed only a small fraction of cosmic history. That's because the human race has been improving so rapidly in knowledge and technology that if people had been around for millions of years, the human race would be much further along in its mastery."

Is this tangent supposed to reveal why the universe was created?

Since Hawking brought it up let's look at other explanations. If a nuclear war takes place and a few thousand people have to start all over again, his logic would not apply. A better hypothesis might state that ancient cultures were destroyed by natural catastrophes like tsunamis, earthquakes, and volcanoes. Or they might have destroyed themselves by their own "advancement." Civilizations have disappeared due to war, depravity, and corruption—like Rome.

According to Plato and his followers, Atlantis was an advanced civilization in antiquity that was wiped out, for some unknown reason. Archeologists have discovered various lost cultures of antiquity underwater. Some examples include the Yonaguni Monument in the Japanese archipelago, the sunken cities below the Gulf of Khambhat and off the coast of Dwaraka in India, as well as the underwater city of Wanaku in Lake Titicaca, Peru.

These are forgotten chapters of the human story, proving our history is not one of uninterrupted progression. The fact that we don't remember simply means we are a species suffering from amnesia.

After the tangent, Hawking returns to the creation myth issue with another jab at people who accept the traditional story. "Bishop Usher [sic], primate of all Ireland from 1625-1656, placed the origin of the world at nine in the morning on October 27, 4004 BC."

Why in the world is this trivia in a serious physics book? Moreover, Hawking is mistaken. Archbishop James Ussher dated the creation of the world as the night prior to Sunday October 23, 4004 BC. Many scholars offered calculations, even Sir Isaac Newton proposed a date circa 4000 BC.

To be fair, Hawking does quote British physicist Sir William Thomson, also known as Lord Kelvin, who wrongly declared in 1884: "One thing we are sure of, and that is the reality and substantiality of the luminiferous ether."

What Kelvin held as knowledge is now considered ignorance. Thus, Dr. Hawking establishes that scientists can also be firm in their convictions which later turn out to be nonsense.

But why is there a digression towards false claims? Is this a portent of what's to come in *The Grand Design*?

In the 1960s the Steady State model of the universe became popular. Today, the public hasn't heard of this model. It's not even taught in schools. The fact is clear, as science marches on previous knowledge turns into ignorance.

At this point, Hawking wants to warn us not to make a mistake.

He insists that it is "wrong to take the Big Bang literally, that is, to think of Einstein's theory as providing a true picture of the *origin* of the universe. That is because general relativity predicts there to be a point in time at which the temperature, density, and curvature of the universe are all infinite, a situation mathematicians call a singularity. To a physicist this means that Einstein's theory breaks down at that point and therefore cannot be used to predict how the universe began, only how it evolved afterward."

This statement contradicts Hawking's PhD thesis which we briefly looked at in Chapter Four. He continues, "So although we can employ the equations of general relativity and our observations of the heavens to learn about the universe at a very young age, it is not correct to carry the Big Bang picture all the way back to the beginning."

Now it's clear. Einstein's theory can't predict the origin of the universe, and therefore Hawking's own PhD thesis is rendered null and void. Undaunted, he labors on.

> "Since we cannot describe creation employing Einstein's theory of general relativity, if we want to describe the origin of the universe, general relativity has to be replaced by a more complete theory. One would expect to need a more complete theory even if general relativity did not break down, because general relativity does not take into account the small-scale

structure of matter, which is governed by quantum theory."

Based on this remark, Hawking now turns to cosmic inflation to try and overcome the biggest hurdle in every origins story, a beginning point, or the beginning of time.

Here is the research on inflation. Theoretical physicist Alan Guth, then of M.I.T., first proposed the idea in the early 1980s suggesting that inflation was due to an anti-gravity force. During the first tiny fraction of a second after the Big Bang, he said, an anti-gravity field caused a runaway expansion of the universe. His hypothesis went beyond Einstein's relativity theory to invoke features of quantum theory.

Later in 1984, Guth and Paul Steinhardt wrote an article in *Scientific American* admitting the limitations of the idea, and that it afforded the opportunity for speculation:

> "The inflationary model of the universe provides a possible mechanism by which the observed universe could have evolved from an infinitesimal region. It is then tempting to go one step further and speculate that the entire universe evolved from literally nothing."[17]

Of course, if "the entire universe evolved from literally nothing" it would mean that "literally nothing" could have both mass and spatial extension.

Physics doesn't have an answer for what happened at the very beginning to create space out of literally nothing, if indeed that was the case. Presumably, a theory of quantum gravity could solve that problem.

By 2001, not much had changed in cosmology. *The New York Times* concluded in its science segment: "The only thing that all the experts agree on is that no idea works—yet."[18]

---

[17] Alan Guth and Paul Steinhardt, "The Inflationary Universe," *Scientific American*, May 1984, p. 128
[18] "Before the Big Bang There Was What?" *The New York Times*, May 22, 2001

Thirty years after Alan Guth, physicists are still not sure how inflation happened. So in three decades little progress has been made. But for all the problems that inflation solved, it created others because it wiped out the possibility of knowing anything that existed previously. So now we have no idea what happened before inflation; we can only speculate.

Another problem: we are told that the inflationary expansion happened so fast that it far exceeded the speed of light, the maximum speed limit of the cosmos. Yet Hawking writes that the speed of light doesn't apply to the expansion of space itself.

This is interesting conjecture, but where is the data? Without proof it's only speculation; what Richard Dawkins calls "blind faith without evidence." We revisit this issue in Chapter Thirteen.

Physics still doesn't have a quantum theory of gravity, just ideas. Yet Hawking will justify his fusion of quantum theory and relativity theory via Alan Guth.

Therefore, he writes "just as we combined quantum theory with general relativity—at least provisionally—to derive the theory of inflation, if we want to go back even further and understand the origin of the universe, we must combine what we know about general relativity with quantum theory."[19]

Did quantum events take place at the onset of the Big Bang? We can take the opinion of other cosmologists into consideration. Professor Paul Davies assures us that "quantum cosmology"—uniting the study of the universe with the study of atomic and subatomic systems—is an ambitious but questionable endeavor. "For a start," he says, "when dealing with quantum gravity, quantum mechanics has to be applied to space-time, not to matter, raising deep technical and conceptual problems."[20]

Again, Hawking doesn't seem to be bothered by the opinion

---

[19] Hawking and Mlodinow, *The Grand Design*, Bantam Books, New York, 2010, p. 207
[20] Paul Davies, *The Goldilocks Enigma*, Penquin Books, London, 2007, p.86

of other physicists. Instead of considering the objections of colleagues, he prefers to go poetic, borrowing the metaphor of a flat world to press his point about the beginning of time using the inflation model.

> "The issue of the beginning of time is a bit like the issue of the edge of the world. When people thought the world was flat, one might have wondered whether the sea poured over its edge."

This quaint idea is an old wife's tale. Pythagoras wrote that Earth was a sphere as early as the sixth century B.C. Later, Aristotle and Euclid also wrote about a spherical Earth. At the height of the Roman Empire, Claudius Ptolemy wrote *Geography* stating that our planet was a sphere.

Even in the early 1200s a book titled *The Sphere* was published about a spherical Earth. By the 1300s this book was required reading in many universities of medieval Europe. It was still in use until the 1700s.

In 1991, University of California professor Jeffrey Burton Russell wrote *Inventing the Flat Earth*. His book clarified how the flat earth idea came to be accepted due to the inaccurate writing of Washington Irving, Antoinne-Jean Letronne, and others during the 1800s.

Moreover, people who lived by the sea sustained themselves by seafood. They saw the boats go out daily and watched them disappear over the horizon. But they always returned with the day's catch.

On the boat, the fishermen would see their village disappear but everything was there as usual at evening. The same with conquering armies that went to sea, and persons who watched vessels sail in regularly with goods from other countries. Nobody thought the sea poured over the edge of the earth because no boat ever approached the edge of the earth.

If a frightened young girl saw father's boat disappear over

the horizon, mother assured her that there's nothing to fear; the boat would return at evening.

So the idea of falling off the edge of the world is a bogus concept. If *The Grand Design* is meant to be a serious book on cosmology, why does Hawking refer to an old wife's tale with zero merit?

Yet he will use this false concept to present his theory about the beginning of time. He explains that time either has a beginning or it goes on forever. Einstein's general theory of relativity doesn't resolve this problem. If time had a beginning how did it start?

The modern origins story says time began with an exploding singularity. But that idea always had awkward problems, so Hawking will now present his new version of events where time behaves like space.

> "Although Einstein's general theory of relativity unified time and space as space-time and involved a certain mixing of space and time, time was still different from space, and either had a beginning and an end or else went on forever. However, once we add the effects of quantum theory to the theory of relativity, in extreme cases warpage can occur to such a great extent that time behaves like another dimension of space."

Factually speaking, Hawking's "extreme" case never happens. However, he proposes that it might have happened at the birth of the cosmos. His premise is that when the universe was at Planck size, a billion-trillion-trillionth of a centimeter, i.e. a singularity, this is the scale where quantum theory can be used to explain events.

> "So though we don't yet have a complete quantum theory of gravity, we do know that the origin of the universe was a quantum event."

He posits that *if* the origin of the cosmos is governed by quantum theory and relativity theory, the result might be that time behaves like space; but only in extreme cases. Thus he speculates:

> "There were effectively four dimensions of space and none of time." He concludes that there is no time at the origin of the universe, because it's just another dimension of space."

This conclusion conveniently avoids the sticky question of how time began by changing its nature into space. But it's reached by fuzzy logic at best. Hawking assumes that the quantum physics of sub-atomic particles with infinitesimal mass, will also work with infinite mass crunched into infinitesimal space. But where is the data to support this assumption? Without data, or even a workable quantum theory of gravity, it's merely conjecture.

Even if it could happen, it could only happen once. So this is not a law, nor is it a regular occurrence in nature. Does anybody accept this as empirical fact? Most people would consider it a miraculous event. In spite of this, Hawking will now accept his idea as given.

As hard as it is to believe, Stephen Hawking and Leonard Mlodinow are prepared to stake their reputation on a miracle event that might have happened once—so it's not a law—if indeed it ever happened at all.

Their position is a bizarre idea, with no data, and a miracle to boot. But science does not support miracles. Ergo, the argument is dead; a still birth argument.

We have the right to ask, if time is postulated to behave like space in the beginning how did it again become like time? No answer is forthcoming.

Yet Hawking assures us: "We must accept that our usual ideas of space and time do not apply to the very early universe.

That is beyond our experience, but not beyond our imagination, or our mathematics."

In Chapter Eight, I will discuss how mathematics can be manipulated to produce practically whatever result one desires. So Hawking's statement is simply his imagination, as he himself implies.

What happens to the beginning of time if all four dimensions behaved like space in the very early cosmos? Here is his answer:

> "The realization that time can behave like another direction of space means one can get rid of the problem of time having a beginning, in a similar way in which we got rid of the edge of the world."

This method of avoiding a sticky question by using an inappropriate argument is clearly fallacious in its reasoning. Let's follow his argument step-by-step:

1. the world having an edge creates a problem
2. but the world doesn't have an edge
3. therefore, the problem disappears
4. time having a beginning creates a problem
5. but time is like a direction of space at the birth of the cosmos
6. therefore, the problem of time disappears

When people abandon logic and common sense they can blindly accept any theory. If we do not abandon logic and common sense we can easily see the error in an argument.

First, there never was a problem with the edge of the world. Nobody fell off the edge of the world and daily experience was the proof. There was no problem to get rid of except, perhaps, in the minds of blind followers.

Second, there is no problem with time. The dilemma is finding a proper explanation how time began *if* the Big Bang model represents reality. I deal with this at length in Chapter Thirteen.

The third problem with Hawking's logic is that time is a completely different concept than space. Einstein fused them together mathematically as space-time. Thus, space-time is simply a mathematical space that is specified by space and time coordinates. To say time behaved *like* space is another kettle of fish entirely. That requires a huge leap of faith; a leap few scientists will stake their reputation upon.

Fourth, at the explosion of the singularity inflation takes over. The universe expands enormously within a split second, well beyond the influence of quantum physics which acts only on atomic scale. So in the initial split second after the Big Bang all quantum forces perish as inflation overrides all laws.

More to the point, it is highly questionable whether quantum forces could even act on the singularity due to its infinite mass. Quantum physics applies to infinitesimal mass.

Earlier in his book Hawking stated that nothing can be said about a beginning using Einstein's equations. Consequently, relativity theory does not apply before the explosion of the Big Bang. So what are we left with?

Cosmology has told us there was a singularity in which infinite density, infinite mass, infinite temperature and curvature, were all squeezed into an infinitesimal size, a Planck size (that is 10^{-20}). An extraordinary situation. The singularity could not contain its cargo, and so it exploded. As soon as the singularity exploded the clock began ticking, according to some.

We may propose that the clock was already ticking prior to the explosion because, clearly, the singularity existed before it burst forth onto the scene. Yet no mention is made of the time factor relating to the existence of that singularity which contained all the matter, energy, and laws of the soon-to-be-born universe. It's similar to a gestation period.

I don't count the nine months being in the womb, because the clock starts as soon as I'm born. However, for my mother, time begins nine months earlier. Scientists may not like this analogy because 1) it tends to lend credence to the traditional creation story, and 2) there is no science that can verify the singularity as a factual object.

However, in order for events to occur, time must already exist prior to, and independent of, an event. Otherwise, time begins the moment an event begins. In either case, it follows that time already existed before the Big Bang because of the assumption of the existence of a singularity, which subsequently became our universe.

Ultimately, Hawking fails to explain the mechanism that merges quantum theory and relativity theory so that they work in tandem. He skirts the issue and tries to assure us that "…inflation has to be there in order for the theory to work, even though the explanation defies all other concepts of physics."

This admission is quite revealing. It has to be that way for the theory to work? Not because the real cosmos is that way? It's like Einstein adding his cosmological constant to fit a prevailing view without the prerequisite data.

Here again is proof that physicists shoehorn their ideas, hopes, and aspirations, to derive a theory even when the explanation "defies all other concepts of physics." By this definition, the theory itself is a miracle.

To make matters worse, Hawking will now offer the reader another imaginative example to buttress the "beginning of time" hypothesis. We will be asked to suppose that the beginning of the universe was like the South Pole of the Earth, with degrees of latitude playing the role of magnitude.

"As one moves north," Hawking explains, "the circles of constant latitude, representing the size of the universe, would expand. The universe would start as a point at the South Pole, but the South Pole is much like any other point. To ask what happened before the beginning of the universe would become a meaningless question, because there is nothing south of the South Pole."

There is nothing south of the South Pole, and this somehow proves we can now dispose of time = zero? Hawking's argument may sound profound sitting at his desk. He may draw a diagram of the Earth and the South Pole appears to be at the bottom and the North Pole at the top. But, if we go to the South Pole and the North Pole we will get a real world experience.

At the North Pole we may think we're at the top of the Earth and we can only move south in every direction. But at the South Pole, the situation is similar. You're not upside down. You might say we can only move north, but factually we can move in whatever direction we want.

"South" is merely an arbitrary direction, a semantic construct invented by man. By direct observation we can move in every direction at the South Pole exactly like at the North Pole, whether we name it south, north, or butterscotch. It makes no difference how we label it.

But let's continue with Hawking's analogy. The degrees of latitude eventually become smaller and end at the North Pole. The universe and time come to an end because there is nothing north of the North Pole?

For the empirical study of the real cosmos, this analogy has no connection to factual reality. It's certainly not physics. "South of the South Pole" is just a semantic riddle that might only have some sense as a brain-teaser. It has nothing to do with the physical reality that physicists study, or even a real life experience at the South Pole.

Clearly, Hawking needs a much better analogy because "south of the South Pole" defeats his own case in terms of reason and logic. It's a matter of semantics, not of physics, because his South Pole argument doesn't hold in terms of physical reality. Nor does its application to the origin of the universe hold in terms of physical reality.

Both analogies in terms of the edge of the Earth, as well as its latitudes, are false. The first equates time and the edge of the earth on the basis that they are both problems, though other than they are a problem they have nothing in common. The second likens latitudes with time, though they also have nothing in common, and since latitudes have limits, time must too if the analogy holds.

His severe weakness in logic and philosophy leads him to erroneous and absurd conclusions that are just words, really, with zero merit for physical reality. And that's the failing of his book so far.

However, Hawking truly believes his argument has merit:

> "The realization that time behaves like space presents a new alternative. It removes the age-old objection to the universe having a beginning, but also means that the beginning of the universe was governed by the laws of science and doesn't need to be set in motion by some god."

Factually, there is no "realization that time behaves like space." It was only Hawking's arbitrary decision to describe it that way. And his description implied a miraculous event that defied all known laws of physics if it even happened at all. So how was it "governed by the laws of science?" We need hard data to substantiate that time behaved like space.

Science can't take credit for laws that existed billions of years before science existed? The laws of nature are universal laws, of which science has only discovered a small percentage. Remember, 96 percent of the energy and matter of the cosmos is unknown. Whatever laws govern this major portion of the cosmos are also unknown.

Once again, Hawking skirts the issue of the origin of the

laws. Could it be south of the South Pole?

By sweeping it under the rug maybe he thinks it won't come up for discussion. This is like a hired housekeeper who accepts money for cleaning but sweeps dirt under the rug thinking no one will notice. It means the house of physics is not clean, at least not Hawking's house of physics where dirt is hidden.

It also indicates he's not being straightforward. This is the impression we are left with. Otherwise, he has a lot of explaining to do and a lot of unanswered questions to deal with. Where did the laws come from? How were they present before the creation of the universe? How were they formulated to govern the cosmos?

Science is still in a fledgling state. But comparable to the "facts" of Bishop Usher and Lord Kelvin, Hawking's thesis may well be a source of jokes in the 22nd century. He denigrated philosophical reasoning at the beginning of his book, but from what we have considered it's now clear his ideas lack logic and scientific rigor to support his case.

If history is correct, and it usually is, we can safely conclude that within the next 50 years science will discover new laws. We'll have more knowledge, move forward, and leave the ignorance of today behind. But until then, today's ignorance will still be accepted as knowledge.

Seventh Chapter

Thus far in this courtroom drama, our attorney has dealt Hawking a serious blow. *The Grand Design* book appears to be an incomplete study of an outdated perspective of 20th century physics, filled with wild speculations and false analogies.

Hawking's next chapter deals with the Goldilocks Enigma and is called "The Apparent Miracle" wherein he examines the bio-friendly fine-tuning of the universe.

We have already discussed the fine-tuning issue in detail in my Chapter Three, so there is no need to go over it again. But we may consider some rhetorical questions raised by Professor Hawking at the end of his sixth chapter.

"What are we to make of this fine-tuning?" he asks. "Is it evidence that the universe, after all, was designed by a benevolent creator? Or does science offer another explanation?"

Well, we've been waiting for another explanation since page one. Hopefully, we'll find that explanation in the final chapter of his book. Thus far, he has not credibly resolved the issue of a beginning for the modern creation story. Will we finally discover his vision for a unifying theory?

Eighth Chapter

This last chapter of Hawking's book is titled "The Grand Design." At last we come to his grand vision of a unifying theory, hopefully. He begins with the following statement: "regularities in the motion of astronomical bodies such as the sun, the moon, and the planets suggested that they were governed by fixed laws..."

Again, the issue of the origin of the laws is sidestepped, but moving on, now he declares that there "must be a complete set of laws that, given the state of the universe at a specific time, would specify how the universe would develop from that time forward. These laws should hold everywhere and at all times; otherwise they wouldn't be laws."

Of course we know that he will present a multiverse conception, with endless universes and infinite sets of laws. This contradicts his definition that a law "should hold everywhere and at all times".

Even in our one universe there are different sets of known laws and not all of them "hold everywhere and at all times". We have quantum physics which only works for sub-atomic particles, and Newton's laws which do not apply to sub-atomic reality.

Undeterred by any sense of contradiction Hawking continues: "At the time that scientific determinism was first proposed, Newton's laws of motion and gravity were the only laws known. We have described how these laws were extended by Einstein in his general theory of relativity, and how other laws were discovered to govern other aspects of the universe."

It looks like we will now get to the essence of Hawking's thesis. "The laws of nature tell us *how* the universe behaves, but they don't answer the *why* questions that we posed at the start of the book.

Why is there something rather than nothing?

Why do we exist?

Why this particular set of laws and not some other?"

My own *why* question is why we had to wade through all the other stuff when we could have gone straight to this immediately? Regrettably, the following explanation provides no answer to the questions:

> "Some would claim the answer to these questions is that there is a God who chose to create the universe that way. It is reasonable to ask who or what created the universe, but if the answer is God, then the question has merely been deflected to that of who created God. In this view it is accepted that some entity exists that needs no creator, and that entity is called God. This is known as the first-cause argument for the existence of God. We claim, however, that it is possible to answer these questions purely within the realm of science, and without invoking any divine beings."

The question of who created God has already been answered,

as Hawking explains, so the problem of deflecting the question does not arise. I'll briefly expand on his explanation.

God is defined as a supernatural entity that exists in a supernatural realm that is eternal. There is no beginning or end to this domain. It is described as supernatural because the effects of time—which we perceive as birth, growth, disease, old age, and death—are absent in that environment. The question of who created God only applies to inhabitants of our physical domain, but makes no sense in an eternal realm with no beginning or end.

However, our interest is in the explanation of the science creation story, so I'm looking forward to that. But, there is yet another digression by Hawking. Instead of answering his own questions quoted above, he refers to the concept of model-dependent realism that he introduced earlier.

> "...our brains interpret the input from our sensory organs by making a model of the outside world. We form mental concepts of our home, trees, other people, the electricity that flows from wall sockets, atoms, molecules, and other universes. These mental concepts are the only reality we can know. There is no model-independent test of reality."

This cries out for a comment. He claims we are dependent on a model to understand the universe because "our brains interpret the input from our sensory organs by making a model of the outside world" which is supposed to prove there is no test for reality independent of a model.

We all see and use such things as wall sockets, homes and so on, plus we can all see the result of experiments that prove the existence of atoms and molecules which allow us a glimpse into their nature. Therefore, the concurring opinion of experts is an indication of the reality of a theory, just as the concurring opinion of people in a room who see a wall socket, is a sure indication that the socket is real.

However, if one person in the room sees a ghost of his grandmother and others do not, the overall conclusion is that it was not real, just a figment of his bereaved mind. Following this logic, we are not required to accept one person's test of reality. When *every* person sees the same thing and describes it the same way, then we may come to the conclusion that it is what it is, beyond our own mental construct, or model. This test of reality is independent of a model.

Next, Hawking makes another false claim: "It follows that a well-constructed model creates a reality of its own." His logic goes like this: Reality is dependent on a well-constructed model. I have such a model. Therefore I have reality.

We have shown that reality can exist independently of an observer. If I don't see it someone else will. In fact, a well-constructed model could just as readily create an *illusion* of its own. The board game "Monopoly" is an illusory model of genuine real estate. How to decipher the real from the illusory? That is the issue.

In this section of his book, Hawking has been resorting to philosophical concepts not scientific methodology. Is this yet another example of hocus-pocus sleight-of-hand to mislead the reader? Why is it difficult for Hawking to give straightforward answers?

If anyone else thinks Professor Hawking is going round in circles, then I'm not alone. Are we finally going to receive the answers from the realm of physics? We've been kept waiting so long that I'm already having doubts.

THE GAME OF LIFE

All of a sudden, the Grand Design becomes a Grand Disappointment. Dr. Hawking is going to model his theory of creation on a board game invented in 1970 by John Conway called *The Game of Life*. Is it because "a well-constructed model creates a reality of its own?"

I won't go into the boring details, but Hawking's model of the universe will mirror the conclusions derived from the rules of the game created by Conway. It sounds unbelievable, yet Hawking admits this in his book so it's not my misconception.

He takes up 13 of the final 19 pages of his book to explain Conway's logic and rules of the game. In these 13 pages he uses the *Game of Life* universe to extract a model for our real-world physical universe. Using the attributes of Conway's model, he extrapolates them to the physical laws of nature.

"As in our universe, in the *Game of Life* your reality depends on the model you employ." As in our universe? This is the fallacy of Begging the Question—the circular argument wherein the conclusion is already in the premise. What is meant to be proven has already been accepted.

Although Hawking accepts his premise as a given, we have shown that a well-constructed model like "Monopoly" could just as easily create an illusion of its own, and often does for the serious player.

> "The example of Conway's Game of Life shows that even a very simple set of laws can produce complex features similar to those of intelligent life. There must be many sets of laws with this property. What picks out the fundamental laws (as opposed to the apparent laws) that govern our universe? As in Conway's universe, the laws of our universe determine the evolution of the system, given the state at any one time. In Conway's world we are the creators—we choose the initial state of the universe by specifying objects and their positions at the start of the game."

A simple point of contention: Hawking can't choose the initial state of the universe (as he can in Conway's game) because the initial state of the cosmos, as well as the laws that govern the universe, were already "chosen" at the Big Bang. Dr. Hawking is a Johnny-come-lately.

The original conditions and the basic laws of physics already limit what physical systems can and cannot do. So my question to Professor Hawking still remains: How did these laws come into being before humans existed?

It makes no difference whether the cosmos began with a Big Bang, a Little Bang, or Many Bangs. Nobody can change the rules of the universe because the laws are pre-determined. We're interested in solving the puzzle of how everything came into being. Hawking promised us this was forthcoming, but he is only extrapolating from Conway's board game.

Why doesn't Hawking come up with original ideas? Always borrowing from others reduces his stature and reputation. Perhaps John Conway should get credit for the Grand Design Theory. Finally, in the last three pages of his book Professor Hawking presents his explanation.

> "In a physical universe, the counterparts of objects such as gliders in the Game of Life are isolated bodies of matter. Any set of laws that describes a continuous world such as our own will have a concept of energy, which is a conserved quantity, meaning it doesn't change in time. The energy of empty space will be a constant, independent of both time and position. One can subtract out this constant vacuum energy by measuring the energy of any volume of space relative to that of the same volume of empty space, so we may as well call the constant zero."

He has arbitrarily assumed that the energy of empty space is a zero constant, yet the presence of dark energy is necessary to explain the expanding universe, even in empty space. What follows is clearly deceptive:

> "If the total energy of the universe must always remain zero, and it costs energy to create a body, how can a whole universe be created from nothing?"

From assuming that the energy of empty space is a zero constant, he concludes that the total energy of the universe "must always remain zero". The erroneous conclusion that now the total energy of the universe must also be zero is clearly misleading.

"If" is a small word with a huge meaning. My simple point is this: nobody can suggest it's actually zero without presenting data as evidence. Yet after arbitrarily making it zero, Hawking accepts it as a given because he believes "a well-constructed model determines a reality of its own."

Physicist J. J. Thomson, who discovered the electron, was very succinct about the role of using mathematics in physics:

> "We have Einstein's space, de Sitter's space, expanding universes, contracting universes, vibrating universes, mysterious universes. In fact the pure mathematician may create universes just by writing down an equation, and indeed if he is an individualist he can have a universe of his own."[21]

Stephen Hawking is indeed an individualist. By assuming first that the energy of empty space is a zero constant, he concludes that the total energy of the universe must also be zero, and therefore composed of positive and negative energies.

Furthermore, in an attempt to make them add up to zero,

---

[21] Ronald W. Clark, *Einstein: The Life and Times*, Avon Books, New York, p. 301

he arbitrarily assigns them positive and negative values. Thus, by playing with the numbers he comes to this conclusion:

> "...there must be a law like gravity. Because gravity is attractive, gravitational energy is negative. One has to work to separate a gravitationally bound system, such as the earth and moon. This negative energy can balance the positive energy needed to create matter."

His assumption that the energy required to produce matter generates an equal and opposite negative energy called gravity, is based on this: "One has to work to separate a gravitationally bound system, such as the earth and moon."

Again, this is deceptive. What keeps the moon and earth separate, though gravitationally bound, is the tangential velocity of the moon, which is the right speed for it to remain in constant orbit. It has nothing to do with the "positive energy needed to create matter".

Moreover, in Hawking's model gravity *already* exists and it can create matter. No inkling from him how gravity came to exist. Not one word about how any laws came to be, nor what sort of force grounds the laws so that they hold in all cases and continue to govern the universe. These questions about the universal laws are left unresolved. It's like the game of "Monopoly" where you already start out rich. He continues:

> "A body such as a star will have more negative gravitational energy, and the smaller it is (the closer the different parts of it are to each other), the greater this negative gravitational energy will be. But before it can become greater than the positive energy of the matter, the star will collapse to a black hole, and black holes have positive energy. That's why empty space is stable. Bodies such as stars or black holes

> cannot just appear out of nothing. But a whole universe can."

Dr. Paul Davies explains why space can't be empty. It is "provided by quantum mechanics, which predicts that even apparently empty space is teeming with virtual particles."[22]

So the nothingness, or the zero constant, that Hawkings bases his model on, doesn't seem to hold water. He claims that stars and black holes have positive energy so they can't appear out of nothing. But the universe they are within, can. Why?

> "Because gravity shapes space and time, it allows space-time to be locally stable and globally unstable."

Hawking claimed earlier that at the beginning of the universe, "there were effectively four dimensions of space and none of time." If there was no time at the beginning (because it behaved like space) there couldn't have been space-time. This is yet another contradiction in his explanation.

Once again, space-time is a mathematical space whose points must be specified by both space and time coordinates. Without specifying time and space coordinates space-time remains hypothetical.

> "On the scale of the entire universe, the positive energy of the matter *can* be balanced by the negative gravitational energy, and so there is no restriction on the creation of whole universes." [his emphasis]

In the above statement he claims it's possible that the energy can cancel out to zero, although no data is given as supporting evidence. It's an arbitrary idea decided by Stephen Hawking who seems to be playing God here. But even if that was the case how does he come to the following unproven conclusion?

---
[22] Paul Davies, *The Goldilocks Enigma*, Penquin Books, London, 2007, p. 166

"Because there is a law like gravity, the universe can and will create itself from nothing..." From his premises that positive and negative energy cancel each other out, if indeed they do, it does not follow that the universe will create itself from nothing with every law perfectly fine tuned for life thrown into the bargain.

To say that something was caused by nothing means that it was self-created by no other thing that existed. If there was one other thing present then it didn't spring from nothing. If positive and negative energies were causing agents, along with gravity, then clearly the universe did not spring from nothingness. That is not a coherent conclusion.

Moreover, my original question is still left unresolved. How did gravity come into being first, before the universe was created?

The logic that the universe can create itself from nothing sounds more unbelievable than having a god as the creator. At least the god conception denotes a creating agency with intelligence. But again, where did the gravity come from to enable the universe to create itself?

The conclusion of his irrational doctrine lies in the following sentence. "Spontaneous creation is the reason there is something rather than nothing, why the universe exists, why we exist."

The illogical idea of self creation is masked by another term "spontaneous creation" which means the same thing—nothing generated everything by itself.

So is there any data to support spontaneous creation (positive and negative energies arising from nothing) as opposed to god creation? Will Stephen Hawking consign the scientific creation story to the same category as the traditional story—blind faith with no evidence?

When there are two alternatives, is there a mechanism, or rationale, for choosing one over the other beyond an arbitrary or whimsical decision?

Here is Einstein's conclusion: "When two theories are available and both are compatible with the given arsenal of facts, then there are no other criteria to prefer one over the other except the intuition of the researcher."[23]

Professor Hawking's book presents no "given arsenal of facts." There are no criteria to support his creation myth over the traditional creation myth. His preference is just an arbitrary choice.

Anticipating a fire storm of outrage, Hawking presents this explanation:

> "Why are the fundamental laws as we have described them? The ultimate theory must be consistent and must predict finite results for quantities that we can measure. We've seen that there must be a law like gravity, and we saw in chapter five that for a theory of gravity to predict finite quantities, the theory must have what is called super-symmetry between the forces of nature and the matter on which they act. M-theory is the most general super-symmetric theory of gravity. For these reasons M-theory is the only candidate for a complete theory of the universe. If it is finite—and this has yet to be proved—it will be a model of a universe that creates itself. We must be part of this universe, because there is no other consistent model."

No comment from me. If what's been presented "has yet to be proved" this is condemnation enough. Scientist Robert Matthews sums it all up:

---

[23] "Induction and Deduction in Physics," Berliner Tageblatt, December 25, 1919. Cited in *The Expanded Quotable Einstein*, p. 237

"Take quantum theory, the laws of the subatomic world. Over the past century it has passed every single test with flying colors, with some predictions vindicated to ten places of decimals. Not surprisingly, physicists claim quantum theory as one of their greatest triumphs. But behind their boasts lies a guilty secret: they haven't the slightest idea why the laws work, or where they come from. All their vaunted equations are just mathematical lash-ups, made out of bits and pieces from other parts of physics whose main justification is that they seem to work."[24]

The Grand Design certainly let us down. Resorting to a board game invented in 1970 doesn't sound like 21st century physics to me.

To accept Hawking's contention that the universe created itself, we should consider whether the universe is tapping into a universal intelligence or some type of cosmic consciousness. Why? Because mathematical laws and laws of nature indicate some sort of underlying intelligence. The principle of entanglement, whereby particles communicate with each other across the universe, indicates some kind of awareness.

It's logical and reasonable to attribute advanced consciousness to laws which describe and define nature. But Hawking wants us to accept that everything arose spontaneously from nothing. He claims a "grand design" yet he doesn't even consider the subject of consciousness in his model.

Let's consider another thought experiment. The nature of my body, from a quantum physics perspective, is almost entirely space. My skeleton is also mostly space. The nature of matter is that even solid rock is mostly space with tiny hubs of

---

[24] Robert Matthews, *New Scientist*, 30, 1, 1999, p. 24

mass/energy we call atoms. The universe is also mostly space with tiny hubs of mass/energy we call stars and planets. Its composition is qualitatively similar to the human body, although quantitatively quite different.

So the human body is a conglomeration of atoms that follows the laws of nature but at the same time follows the dictates of a conscious will to serve a specific purpose. The question whether there is a will in the universe, could be evidenced by a purpose coming about in the universe.

Since the universe is so bio-friendly one obvious purpose, universally, could be the creation of life. From the definition that my body is simply a conglomeration of atoms pervaded by consciousness, we get the idea that the universe is also a body of matter that might be pervaded by consciousness, perhaps even directed by will.

Because we can't see cosmic consciousness does not mean it's not there, any more than not being able to see my consciousness doesn't mean that I don't exist. The fact that I have a body which serves my purpose is proof that I exist. The idea of a cosmic consciousness might be how most people define God, which could lead to a synthesis of science and theology.

If no link exists between life, consciousness, and the cosmos, why are we discovering laws that work in harmony to produce life plus everything required to support life? Why are scientists searching for a link? Or, shall we conclude that the origin of the laws is random nuclear fusion?

If physicists accept the challenge of investigating the role of consciousness many troubling issues might be better understood. This idea is exactly what the original quantum physicists realized a hundred years ago.

From the perspective of Max Planck, much of the twentieth century was wasted because later quantum theorists neglected to expand on his work. Considered the father of quantum physics, his work demonstrated that consciousness affects the results of quantum physics experiments. In 1918, Planck received the Nobel Prize for his contribution to science.

After a lifetime dedicated to physics, Planck's work convinced him that consciousness was a primary and pre-existing force. He explained it like this:

> "As a man who has devoted his whole life to the most clear headed science, to the study of matter, I can tell you as a result of my research about atoms this much: There is no matter as such. All matter originates and exists only by virtue of a force which brings the particles of an atom to vibration and holds this most minute solar system of the atom together. We must assume behind this force the existence of a conscious and intelligent mind. This mind is the matrix of all matter."

Is Planck's version on the right track? Consider one of the fundamental principles of classical physics—that a change can only come about by an applied force.

For example, the theory of evolution posits that things change and evolve. In order to shift from a settled state there must be some impetus to launch the change. According to the prevailing theory, survival is the motivating stimulus that pushes evolution. Adaptation provides the momentum for change that ensures survival. In this way the forward motion of evolution persists by a force—the conscious will to survive.

Likewise, without some sort of stimulus nothingness would remain as it is. How would the universe come into being from nothing unless an applied force acted upon it? Stephen Hawking's explanation that positive and negative energies supplied some stimulus, contradicts the entire basis and definition of nothingness.

Further, why would a singularity in a state of rest suddenly release its contents, thereby creating the universe? Some impetus at time t=0 must have been the force to provoke the change according to the laws of physics.

The traditional story proposes an impetus from God which begins the creation. This idea remains unproven; however, it does provide an active force to answer the dilemma. Cosmology's failure to provide an explanation puts it at a distinct disadvantage.

Moreover, Hawking's explanation for a "grand design theory" cries out for a response about consciousness, but he doesn't acknowledge it. He ends his book with this comment:

> "M-theory is the unified theory Einstein was hoping to find. The fact that we human beings —who are ourselves mere collections of fundamental particles of nature—have been able to come this close to an understanding of the laws governing us and our universe is a great triumph. But perhaps the true miracle is that abstract considerations of logic lead to a unique theory that predicts and describes a vast universe full of the amazing variety that we see. If the theory is confirmed by observation, it will be the successful conclusion of a search going back more than 3,000 years. We will have found the grand design."

Did you notice the disclaimer? "If the theory is confirmed by observation" tells us that there is no evidence, no data.

Remember what Hawking stated in his third chapter. "According to model-dependent realism it is pointless to ask whether a model is real, only whether it agrees with observation."

If the reader assumed that the Grand Design model *would* agree with observation, it's another disappointment.

Factually, there is zero observational data to support his "spontaneous creation" model. Therefore, by his own statement, its validity is flawed. Consequently, the grand design theory is mostly conjecture and barely even science.

Our lawyer comments that it's now clear why Roger Penrose said Hawking's M-theory was hardly science—"It's a collection of ideas, hopes, aspirations; it's not even a theory." Case closed; game over.

8
Fuzzy Math and Faulty Theories

If you're following the trial, at this point you are probably dismayed by what we have discovered. We always believed that science dealt with facts which we could trust. Yet, a prominent cosmologist like Stephen Hawking presents information that is as speculative as any tribal creation myth.

I now understand that Hawking's brilliance is not in his power of logic or reason. Rather, it's that he has been able to bring physics to the masses. His style is concise and understandable. He adds a dash of humor which I think many appreciate, though some might consider corny.

We also have to acknowledge his determination to stay alive and continue his work after he was written off by doctors that he would never live to finish his PhD. But it's quite disconcerting that an investigative journalist can dismantle a theory of cosmology by a person acclaimed to be brilliant.

Of course, all credit must go to the scientists offering their testimony. They work hard to unravel problems and resolve contradictions. Brave physicists put their career at risk by speaking up when they find data that contradicts the *status quo*.

CRITICAL ANALYSIS

Professor Hawking's book, *The Grand Design*, seems to be his swan song—an attempt to be recognized for providing a unique contribution. However, the book spoils his aspiration to be recognized by history as a great contributor to the field of cosmology.

Instead of giving us solid physics, his book simply served up conjecture and speculation. No grand design of the universe shone forth. No brilliant physics except for a rehash of what Feynman, Guth, Einstein, and even Conway presented years ago. There's nothing new for the cosmology story that hasn't been discussed before, in greater detail, by other physicists.

We may question whether Hawking's conclusions are even original. For example, in the March 1984 edition of *New Scientist* magazine an article by physicist Edward P. Tryon was titled, "What Made the World?"

On page 15 he wrote: "I proposed that our Universe had been created spontaneously from nothing, as a result of the established principles of physics." Tryon says he originally proposed this in 1973.

We already quoted Alan Guth and Paul Steinhardt who later echoed the same idea as Tryon, and now Hawking. So his conception is not original.

Historically, the idea of the self-creation of matter espoused by Hawking's grand design theory has been part of an underlying direction of modern cosmology since the time of Charles Darwin.

Let's suppose the universe could spontaneously erupt from literally nothing. Then "literally nothing" can extend itself into space, mass, energy, gravity, consciousness, love, and infinite varieties of life. Such reasoning suggests that ideas in physics and cosmology are deteriorating into utter nonsense.

Alternatively, we can construe a more reasonable explanation by way of a higher dimensional reality. The so-called "nothing" could not be comprehended simply due to a lack of knowledge, which leads us to an incorrect conclusion: that which can't be detected or figured out is taken to be nothing.

So far we have seen that most of Hawking's book is philosophically, or theoretically, based. What remains is conjecture. There is no observation, no experimentation, and no objective testing to verify the proposed "Grand Design" theory. Because he has no empirical data he resorts to philosophy throughout his book, but he does so by appealing to the use of misleading and inaccurate metaphors.

By saying that the universe can spontaneously arise from nothing Hawking misuses the meaning of "nothing" which is the absence of anything at all. Again, nothing means that "no thing" exists to have any properties that could generate any cause or effect. So nothingness could never be the source of anything, what to speak of the entire universe.

Nor is empty space consistent with nothingness, because space is no longer considered to be empty. The quantum vacuum is also not "nothing"—it is a sea of fluctuating energy.

Because he uses the word "nothing" in an inaccurate sense philosophically, Hawking comes to an erroneous conclusion. And it's hard for me to believe that he has not made this connection himself.

We have cause to question: did he deliberately choose to misuse the meaning of "nothing" to justify his hypothesis?

His book relies upon grandiose and baseless philosophical pronouncements devoid of scientific rigor. The methodology of science requires carefully testing ideas encompassed in a theory. Every hypothesis must be tested rigorously, including attempts to disprove it.

Challenges, critiques, and testing of hypotheses are integral factors of genuine science. When ideas withstand challenges based on known scientific principles and not proven incorrect, then they can be elevated to the status of a theory. When there is no observational data there is no theory, there is just an idea.

Why didn't Hawking give rigorous proofs and substantiate M-theory through logical hypotheses? He could have included research from Ed Witten. He should have included perspectives from other theoretical physicists in the field. But he didn't. So we took him to the court of public opinion.

Physicist Paul Davies is quite frank about the relevance of M-theory: "...it uses branches of mathematics which are not only extremely abstract, but also extremely obscure. In fact, some of the mathematics had to be invented along the way."[25]

The obscure and invented mathematics leaves M-theorists with no reality check. It reminds me of the ugly stepsister trying to fit her big foot into Cinderella's delicate glass slipper.

Professor Hawking concludes that no god is necessary for the universe to have come into being and to exist. Instead, gravity and the laws of nature are necessary for the universe to come into being and exist. But without resolving the enigma of the origin of gravity, the origin of the laws, and why they always work, he leaves a huge gap unexplained.

A theory can't have such a major flaw, a sizable gap left unresolved and unaccounted for, and still maintain its credibility.

Another problem is his statement about the positive and negative energy of the universe canceling out. "On the scale of the entire universe, the positive energy of the matter *can* be balanced by the negative gravitational energy, and so there is no restriction on the creation of whole universes." [his emphasis]

Does he mean that today's matter and gravity creates new universes? Wasn't his grand design theory intended to resolve the initial creation issue? Or, does the positive energy of matter and the negative energy of gravity exist before the cosmos is created? If so, then how are those nothing? This is a huge gaffe.

---

[25] Paul Davies, *The Goldilocks Enigma*, Penquin Books, London, 2007, p. 130

Does Professor Hawking even review what he has written?

Remember, he is trying to solve the issue of the initial beginning of creation when there was nothing in existence—not time, not space, not matter, not energy, and therefore no energies that can oppose each other.

The stumbling block of Hawking's self-creation of universes is that matter, energy, gravity, and M-theory suddenly appeared from nothing and were already present to allow for spontaneous creation. But how did they come into being, and where did they come from? That vital question is never addressed.

Furthermore, if we accept the universe came into being from nothing that begs the question: why don't other things do the same? We should be able to see other things simply appear from nothing if this is part of naturalism. There is no experience of anything ever arising from nothingness, yet Dr. Hawking claims this for an extreme case, the universe itself. But that can never be verified or falsified. It can only be accepted on blind faith.

The bio-friendly nature of the cosmos is another major problem for Hawking. He explains it away with unlimited roulette wheels; the multiverse idea. But it's not convincing. The multiverse can never be verified or falsified because it lies outside of our observable universe. This idea is similar to the unseen realms that religions describe which also lie outside our observable universe.

This makes his book look like an amateur attempt at scientific philosophy; or an abuse of philosophy. It's like a manual for the professional gambler: "bet on multiple wheels to improve your chances." Needless to say, such desperate grasping at straws falls far short of the renowned classical intellects enshrined in the halls of science.

In a *Scientific American* editorial Philip and Phylis Morrison present a brief history of the hoard of cosmological theories that were later overturned.

"We simply do not know our cosmic origins; intriguing alternatives abound, but none yet compel. We do not know the details of inflation, nor what came before, nor the nature of the dark, unseen material, nor the nature of the repulsive forces that dilute gravity. The book of the cosmos is still open. Note carefully: we no longer see a Big Bang as a direct solution. Inflation erases evidence of past space, time, and matter. The beginning—if any—is still unread. It is deceptive to maintain so long the very term that stood for a beginning out of nothing. The chanteuse will compose a clever new song once the case is clear."[26]

An observation by Mark Twain appears appropriate for the grand design theory: "There is something fascinating about science. One gets such wholesale returns of conjecture out of such a trifling investment of fact."[27]

Carl Sagan, in a more conciliatory mood, remarks that, "There are many hypotheses in science which are wrong. That's perfectly all right; they're the aperture to finding out what's right."

Science philosopher Karl Popper summed it up emphatically by stating that, "A scientific theory neither explains nor describes the world; it is nothing but an instrument."[28]

Professor Hawking is a person striving to say something brilliant, but comes up short. He really wants us to accept that his M-theory explanation is "the *only* candidate for a complete theory of the universe." [his emphasis]

"Only" means everything else is rejected. There can be no

---

[26] Philip and Phylis Morrison, "The Big Bang: Wit or Wisdom?" *Scientific American*, February 2001, p. 95
[27] Mark Twain, *Life on the Mississippi*, 1883, p. 156
[28] Karl Popper, Conjectures and Refutations: The Growth of Scientific Knowledge, p. 102

other candidate. Yet his grand design theory is full of conjecture and contradiction. It's unobservable, untestable, and *only* a grand delusion. [my emphasis]

Hawking has failed to see that his creation account is as indeterminate as any traditional creation story. In every creation story the fundamental question is, what started it all? This enigma is also at the root of Hawking's theory. The traditional story claims that a supernatural God of unlimited intelligence and power started it all.

There is no empirical evidence to support the assertion that a conscious being far beyond human intelligence and ingenuity could have brought it all together and set it in motion. Yet, taken at face value, the explanation itself is reasonable enough.

We know that intelligence and ingenuity are required to create and set into motion anything humans are able to accomplish. Based on experience, it's a plausible step to attribute extreme fine-tuning for life to a highly advanced intelligence from another dimension of reality.

Of course, this does not attempt to prove or substantiate any particular god of any particular religion. It only points to advanced intelligence that could have existed prior to life as we know it. After all, the plausibility of consciousness being on a cosmic scale, based on conclusions from the double-slit experiment and the phenomenon of particle entanglement, has already been noted by physicists such as Wigner and Planck.

In a different sense, Richard Dawkins alludes to this in a filmed interview: "It could be that at some earlier time, somewhere in the universe, a civilization evolved by probably some kind of Darwinian means, to a very, very high level of technology, and designed a form of life that they seeded on to, perhaps, this planet. That is a possibility and an intriguing possibility."

Panspermia is the science that considers life on Earth was imported via meteorites, comets, or asteroids. Dawkins continues to support this idea.

"I suppose it is possible that you might find evidence for that. If you look at the D cells of chemistry or molecular biology, you might find a signature of some sort of designer, and that designer could well be a higher intelligence from elsewhere in the universe."

So an advanced consciousness life form is a viable possibility within our vast universe. Dawkins concludes: "But that higher intelligence itself would have had to have come about by some explicable, or ultimately explicable process. It couldn't have just jumped into existence spontaneously. That's the point."

That is indeed the point. For Stephen Hawking to present that the universe just jumped into existence spontaneously from nothing, is not an explicable or plausible proposal. His grand design theory is an example of faith with not a shred of data. We know this is the definition of Scientism.

We don't have to accept Hawking's version blindly and just surrender our native intelligence, sound judgment, logic and reason. A more sensible approach is to take the conclusions of other expert physicists into account.

After reading *The Grand Design*, one can't help thinking about Andersen's tale of the vain Emperor. The grand design theory turned out to be a grand deception.

Hawking and Mlodinow appeared less like scientists and more like swindlers trying to sell us the Emperor's new cosmos.

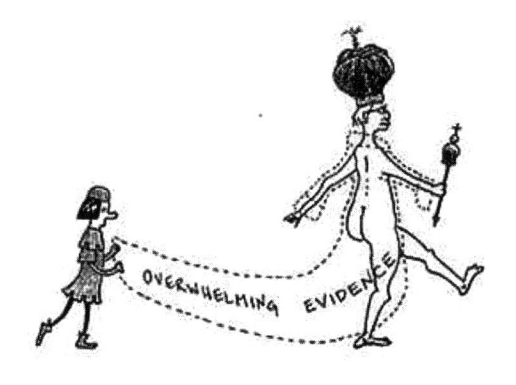

Credit: anonymous political cartoonist emperorswithoutclothes.com

Our attorney will now introduce other scientists to testify that the scientific creation story is not as certain as it's taught. Of course, the purpose of this book is to inform the public of the anomalies in the modern creation story. It's not intended to be a complete treatise on the state of cosmology.

The footnotes in each chapter comprise a bibliography that directs inquisitive readers to the writings of the luminaries of physics. For an in-depth study it's best to read directly from them.

Physics in 1912

The trial continues as our lawyer turns his attention to examine several bedrock concepts of science under the microscope.

One hundred years ago, the modern creation story was much different. What we know today as the Milky Way Galaxy, was then taught in the best universities to be the entire universe! Moreover, the cosmos was believed to be eternal and static. This was the teaching for over three centuries, accepted by all renowned scientists including Einstein.

Looking back at their understanding from our present perspective it appears quaint, even humorous. As we discover more of nature's secrets, the universe has become confusingly complex. Honest scientists willingly admit that the more data our instruments supply the more difficult it is to make sense of it.

Astronomer Fred Hoyle commented that, "The whole history of science shows that each generation finds the universe to be stranger than the preceding generation ever conceived it to be."[29]

A major discovery was the knowledge that the universe operated within strict boundaries that could be understood through mathematics. As Galileo put it: "The great book of na-

---

[29] Fred Hoyle, *Astronomy and Cosmology*, San Francisco, W H Freeman and Co, 1975, p. 48

Fuzzy Math and Faulty Theories

ture can be read only by those who know the language in which it was written. And this language is mathematics."[30]

The British astronomer Sir James Jeans stated it more elegantly. "The universe appears to have been designed by a pure mathematician."[31]

Have we discovered an invisible sub-text of nature? Is there a cosmic code by which the universe obeys mathematical logic? If mathematics is the language of the universe, then how did that language come into existence? Or, have we invented it?

Historically, Sir Isaac Newton did invent the mathematics he needed to describe the laws he discovered. He called his math the theory of fluxions. Today it's known as calculus. He accepted it as an underlying fundamental principle by which the universe was governed—the key to a cosmic code.

Keep in mind that Newton did not have an explanation for the causes of gravity, inertia, and centrifugal force, for which he became famous. But he did have a talent for figuring out the mathematical relationships of these mysterious forces.

Since then, physicists have used mathematics to interpret the Laws of Nature and predict the outcome of natural phenomena.

The Math of Physics

Today, theoretical physicists create mathematical formulations to model real world situations. They fiddle with figures in mathematical equations. By manipulating the numbers, they try to figure out what would happen in real world conditions without performing the experiment. They work in an office, not a laboratory. In this way they're like armchair pilots with flight simulators.

They predict various outcomes and look for evidence to

---

[30] Paul Davies, *The Goldilocks Enigma*, Penquin Books, London, 2007, p. 9
[31] James Jeans, *The Mysterious Universe*, Cambridge University Press, 1930, p.140

support their ideas. When the ideas are not confirmed, they're just dropped. But when they do find something to confirm an idea they say the theory predicted it.

Thus, using a successful theory, a physicist can predict an answer for a problem by simply applying the relevant mathematical equations that model the laws under examination. And that's why they say the theory predicted the outcome. But, as we shall see, to get the desired result the numbers are manipulated in diverse ways, producing various results, some acceptable and others nonsense.

If a mathematical formula accurately simulates a known law, then a physicist can calculate on paper what *might* happen in real life. Does this method work in all cases? And what if the equation is not an accurate representation?

Mario Livio, head of the science division of the Hubble Space Telescope, explains how mathematics simply fills a role that cosmologists need to describe events and concepts.

> "The success of pure mathematics turned into applied mathematics, in this picture, merely reflects an overproduction of concepts, from which physics has selected the most adequate for its needs—a true survival of the fittest. After all, "inventionists" would point out, Godfrey H. Hardy was always proud of having "never done anything useful." This opinion of mathematics is apparently espoused also by Marilyn vos Savant, the World Record holder in IQ—an incredible 228. She is quoted as having said, "I'm beginning to think simply that mathematics can be invented to describe anything, and matter is no exception."[32]

---

[32] Mario Livio, *The Golden Ratio*, New York, Random House, 2002, p. 245

Savant, an American author, lecturer, and playwright, is listed in the *Guinness Book of World Records* under "Highest IQ".

Godfrey Hardy was a prominent English mathematician known for achievements in number theory and mathematical analysis. So these opinions come from renowned experts.

We can question why nature is understood by mathematics. Why do the equations actually work? The fact that nature seems to conform efficiently to mathematical formulations is left unexplained.

Joseph Needham points out that mathematics had already encroached upon physics by the turn of the 20th century. It was invented to satisfy the requirements of the day.

> "The mathematisation of physics…is continually growing and physics is becoming more and more dependent upon the fate of mathematics.…This special mathematics has for the greater part been created by the physicists themselves, for ordinary mathematics is unable to satisfy the requirements of present day physics."[33]

Physicist Herbert Dingle is quite clear that mathematics can substantiate whatever is required.

> "…in the language of mathematics we can tell lies as well as truths, and within the scope of mathematics itself there is no possible way of telling one from the other. We can distinguish them only by experience or by reasoning outside the mathematics, applied to the possible relation between the mathematical solution and its supposed physical correlate."[34]

The science philosopher, Karl Popper, establishes that physics

---

[33] *Science at the Crossroads*, "Marx's Theory on the Historical Process," London, Frank Cass and Co., 1971, p. 189
[34] *Science at the Crossroads*, London, Frank Cass and Co., 1971, p. 33

by mathematics does not correlate to real world situations.

> "Properly understood, a mathematical hypothesis does not claim that anything exists in nature which corresponds to it...It erects, as it were, a fictitious mathematical world behind that of appearance, but without the claim that this world exists. [It is] to be regarded only as a mathematical hypothesis, and not as anything really existing in nature."[35]

Does the mathematics of Einstein represent what really exists in nature? Or does the math merely save appearances? For example, in Relativity Theory is it just a case of saying 4 + 4 = 8 when in reality the correct equation is 5 + 3 = 8?

Morris Kline is a professor of mathematics at New York University and the Courant Institute. He is highly critical of mathematics and its applications to science. Here is the shocking analysis from an insider.

> "It is now apparent that the concept of a universally accepted, infallible body of reasoning —the majestic mathematics of 1800 and the pride of man—is a grand illusion. Uncertainty and doubt concerning the future of mathematics have replaced the certainties and complacency of the past.
>
> "The disagreements about the foundations of the 'most certain' science are both surprising and, to put it mildly, disconcerting. The present state of mathematics is a mockery of the hitherto deep-rooted and widely reputed truth and logical perfection of mathematics. The disagreements concerning what correct mathematics is and the va-

---

[35] Karl Popper, *Conjectures and Refutations: The Growth of Scientific Knowledge*, p. 102

Fuzzy Math and Faulty Theories

riety of differing foundations affect seriously not only mathematics proper but most vitally physical science."[36]

I'm shocked that mathematics can be manipulated according to the subjective intention of a theorist, thus rendering it unreliable and in some cases completely suspect.

Thomas Van Flandern points out that, "Mathematics should be used to describe the operation of models, not to build them."

> "...equations cannot be made to substitute for the concepts which underlie them. And equations are generally blind to limitations of range and physical constraints. They are too general, and simply lack the sort of specificity that true, intuitive understanding demands. Every equation has a domain of applicability—usually the range of the observations and little, if anything, more... If an equation can be extrapolated outside its domain and gives a singularity (basically, a zero divisor), that singularity does not exist in nature; instead, the model needs modification. Up to now this rule has always proved true. But advocates of "black holes" in the universe would have us believe that the equations which predict them can be relied upon far outside the domain of the observations used to derive those equations."[37]

The mathematical singularity does not exist in nature? Physicist Michael Duff is quite candid about this:

> "Well, the question we often ask ourselves as we work through our equations is: 'Is this just

---

[36] Morris Kline, *Mathematics: The Loss of Certainty*, Oxford University Press, 1980, p. 6
[37] Tom van Flandern, *Dark Matter, Missing Planets and New Comets,* revised edition, Berkeley, CA: North Atlantic Books, 1993, p. xxi

fancy mathematics, or is it describing the real world?'"[38]

In his comments on the issue, the celebrated British philosopher Bertrand Russell is particularly sarcastic.

> "Pure mathematics consists entirely of assertions to the effect that if such and such a proposition is true of anything then such and such another proposition is true of that thing. It is essential not to discuss whether the first proposition is really true, and not to mention what the anything is, of which it is supposed to be true. Both of these points would belong to applied mathematics... Thus mathematics may be defined as the subject in which we never know what we are talking about, nor what we are saying is true."[39]

Dr. Lee Smolin represents the Perimeter Institute for Theoretical Physics. He claims that the mathematisation of physics has resulted in the reduction of the cosmos to a mathematical entity. This has not only confused physicists but accounts for their worst and most distracting assertions.

Current speculations about a theory of everything have pushed science beyond the wildest imagination of science fiction. Is M-theory a mirage of mathematics, or can it lead to an ultimate theory of everything?

Mathematicians have now informed us that the use and function of mathematics beyond its traditional scope of explaining observations, has evolved into creating models that have no basis in observation of the real world. This calls the viewpoints of physicists into question.

Even the areas of physics that the public believes are well

---

[38] "A Conversation with Brian Greene," Nova television series, PBS, October 2004
[39] Bertrand Russell, *Mysticism and Logic*, Doubleday, 1957, pp. 70-71

understood by the scientific community, are gradually being revealed as clouded with doubts and disagreements.

Most people believe that with all our scientific knowledge we know what causes the most common occurrence in the world—gravity. Is this a fact? Has modern science understood the cause and source of gravity?

Gravity

During this second week of cross examination, our attorney will now shift his attention from the frailties of Stephen Hawking's grand design hypothesis. His intention is to explore the mystery of gravity to determine if science has actually understood what causes gravitational force.

As he begins his presentation, he turns to the media people present in court and remarks, "Fasten your seat belts."

Gravitational force is responsible for keeping our feet on the ground. It keeps our planet revolving around the Sun. We tend to think of gravity as a strong force because it holds the entire solar system together. Yet scientists describe gravity as a weak force compared to the other known forces.

For example, a small refrigerator magnet can pick up a pin off a table. Physicists cite this event to illustrate how feeble gravity is compared to the magnetic force of an ordinary fridge magnet.

Lisa Randall of Harvard University tried to expound on the phenomenon of gravity for the BBC Documentary: *Gravity— The Weak Force.*

> "There are various forces that we see in nature. Most of them we understand at some level. And then there's gravity, which seems very different. The gravitational force is extremely weak in comparison with the other forces."

This may sound peculiar to most of us. "But if you think about it," Randall continues, "you have the entire earth pulling on you, and yet you can manage to pick things up."

Physics has tried to describe gravity for three centuries. It's portrayed as an attractive force, but that's ostensibly to satisfy appearances. Sir Isaac Newton is credited with discovering the force of gravity, yet he only gave a mathematical formula to calculate its effects.

We've all heard the story: an apple fell and Newton had an epiphany realizing that some sort of force was at work. Was it a force inherent in matter that attracted the apple to the earth, or was it a downward force pushing from above?

Although he was able to calculate the rate at which the apple fell, Newton never claimed to know the cause. But he could scarcely believe gravitational force to be something inherent in matter. Rather he thought that gravity was caused by some external force.

> "That gravity should be innate, inherent and essential to matter, so that one body may act upon another at a distance through a vacuum, without mediation of anything else, by and through which their action and force may be conveyed from one to another, is to me so great an absurdity that I believe no man who has in philosophical matters a competent faculty of thinking can ever fall into it. Gravity must be caused by an agent acting constantly according to certain laws; but whether the agent be material or immaterial I have left to the consideration of my readers."[40]

Well, Newton didn't mince words. To define gravity as a local

---

[40] Letter to Richard Bentley, February 25, 1693, *Newton's Correspondence*, registered in the Royal Society in 1675, Correspondence, vol. 3, p. 253

Fuzzy Math and Faulty Theories

force due to some innate quality of matter does create problems. It doesn't explain how gravitational force operates over vast distances as if transported by some mysterious and unexplained means. This is the "action-at-a-distance" problem that Newton never resolved.

Arthur Koestler writes extensively on universal law, our perception of the cosmos, and gravity. He compares the study of gravitational force to walking in a minefield.

> "Newton avoided the booby traps strewn over the field: magnetism, circular inertia, Galileo's tides, Kepler's sweeping-brooms, Descartes' vortices—and at the same time knowingly walked into what looked like the deadliest trap of all: action-at-a-distance, ubiquitous, pervading the entire universe like the presence of the Holy Ghost. The enormity of this step can be vividly illustrated by the fact that a steel cable of a thickness equaling the diameter of the Earth would not be strong enough to hold the Earth in its orbit."[41]

Our planet travels at a velocity of approximately 30 km/sec (67,500 mph) relative to the Sun. Science tells us that gravity holds it in orbit. When an object spins around, it's trying to escape circular motion due to its tangential velocity. So there must be centripetal force pulling the object towards the center to hold it in place. Otherwise it would fly off into space.

At 30 km/sec, according to Koestler's calculations, a steel cable eight thousand miles thick would not be strong enough to counteract the tangential velocity of the Earth and hold it in orbit. Yet a tiny fridge magnet defies gravity and doesn't fall.

To show that Newton's equations of gravitational force are

---

[41] Arthur Koestler, *The Sleepwalkers : A History of Man's Changing Vision of the Universe*, Pelican Books Ltd., England, 1959, reprinted 1979, p. 344, p. 511

vulnerable, Koestler applies the following logic: Consider the distance between the Earth and the Sun. It takes sunlight 8.5 minutes to reach us. If the Sun suddenly stopped shining we would only detect the absence of light 8.5 minutes later. Of course, we don't notice the travel time because sunlight is discharged constantly.

The force of gravity is also constant. Does gravity work in the same manner as light?

Modern science teaches that Earth is held in orbit by the Sun's gravitational force. The force of gravity travels from Sun to Earth in 8.5 minutes at least, but no less if the speed of light is the maximum speed in the universe. If this is the true speed of gravitational force there's no problem, because the gravity connection between Sun and Earth has been established and undisturbed since time immemorial.

But suppose the gravitational force between the Sun and Earth suddenly stopped. In Newton's model, gravity propagates instantly, so our planet would fly off its orbit into deep space immediately due to its tangential velocity, just like a ball being twirled in a circle when the string is cut.

The problem is: what principle of physics accounts for the *immediate* reaction of a planet to fly off into space when the gravitational force is cut? Would there be a delay of 8.5 minutes like with sunlight? Or, does gravitational force travel faster than light?

Well, the speed of gravity has never been measured directly in the laboratory for several reasons. The first is that the gravitational interaction is much too weak. Another reason is that the experiment is still beyond our technological ability. Therefore, the speed of gravity can only be deduced from astronomical observations. Besides, the answer will depend on which gravitational model a researcher employs to describe his observations.

Fuzzy Math and Faulty Theories

Koestler is quite disturbed that physics hasn't resolved how the most fundamental force of nature behaves.

> "The whole notion of a 'force' which acts instantly at a distance without an intermediary agent, which traverses the vastest distances in zero seconds, and pulls at immense stellar objects with ubiquitous ghost-fingers—the whole idea is so mystical and 'unscientific'—that 'modern' minds like Kepler, Galileo, and Descartes, who were fighting to break loose from Aristotelian animism, would instinctively tend to reject it as a relapse into the past...
>
> "What made Newton's postulate nevertheless a modern Law of Nature, was his mathematical formulation of the mysterious entity to which it referred."[42]

Again we see that mathematics lends a feeling of reality to concepts that would simply be mere conjecture.

Koestler is just one among many scientists who say that science is not always on the mark. He's upset that scientific terms like 'universal gravity' and 'electromagnetic field' have become, "verbal fetishes which hypnotized it into quiescence, disguising the fact that they are metaphysical concepts dressed in the mathematical language of physics."[43]

A metaphysical concept refers to a supernatural event which is quite unacceptable to science.

Einstein proposed that nothing can go faster than the speed of light. This caused a problem for Newton's concept of gravity. Thus, a completely different perspective of gravity was introduced in the 20th century. Einstein described gravitational force in relation to a curvature of space. Does Einstein's theory

---

[42] *The Sleepwalkers*, Pelican Books Ltd., England, 1959, reprinted 1979, p. 344
[43] *The Sleepwalkers*, Pelican Books Ltd., England, 1959, reprinted 1979, p. 508

fare any better in explaining gravity?

Space became a fabric that could be warped and stretched by heavy objects like planets and stars. The warping or curving of space creates what we feel as gravity. So our Earth is held in orbit because it follows curves in the fabric of space caused by the Sun's presence.

But heaviness implies an *already* existing force pulling on an object. Physicist Tom Van Flandern has detected several problems with Einstein's gravity hypothesis. Einstein's idea is that space is curved, so Van Flandern asks, "Why would curvature of the manifold even have a sense of 'down' unless some force such as gravity already existed? Logically, the small particle at rest on a curved manifold would have no reason to end its rest unless a force acted on it."[44]

We all know about the weightlessness of space. There is no light or heavy, up or down. The idea of 'down' only exists because the force of gravity is already present.

Van Flandern also notes that "it is not widely appreciated that this is a purely mathematical model lacking a physical mechanism to initiate motion."

Einstein was convinced that nothing could travel faster than light, so he tried to resolve the action-at-a-distance dilemma this way. If the Sun were to suddenly disappear, the gravitational disturbance would cause a ripple effect across the fabric of space. Just like a stone dropped into a pond causes a ripple effect along the surface of the water to the edge of the pond.

Therefore, a gravitational wave would travel through the fabric of space towards Earth. We would not experience the absence of gravity until the wave reached us. Einstein calculated that gravitational waves travel at the same speed as light, but no faster.

---

[44] T. Van Flandern, "Gravity" in *Pushing Gravity*, Matthew R. Edwards, Montreal: C. Roy Keys Inc, 2002, p. 94

But Van Flandern's question remains: how does the motion of a body begin? Why would any particle at rest near a source mass begin to move toward that source mass?

To answer that difficulty, General Relativity posits the presence of gravitational fields. These fields act like some agency passing between source and target. They are meant to convey an action.

Based on this response, Van Flandern counters with the logic that the gravitational fields are dependent on the principle of causality.

> "...all existing experimental evidence requires the action of fields to be conveyed much faster than light speed. This situation is ironic because the reason why the geometric interpretation gained ascendancy over the field interpretation is that the implied faster-than-light action of fields appeared to allow causality violations...Yet the field interpretation of General Relativity requires faster than light propagation. So if Special Relativity were a correct model of reality, the field interpretation would violate the causality principle, which is why it fell from popularity."[45]

Causality violations imply that if speeds faster than light are possible we would be able to go back in time and perhaps kill our father so that our own life could never be caused. So much for that idea!

If we could travel faster than the speed of light we could reach the position of an event before it took place and influence it. Even a particle or a force would influence events that have already happened, in effect, turning back the clock.

Gravity has always been a sticking point for both physical

---

[45] T. Van Flandern, "Gravity" in *Pushing Gravity*, C. Roy Keys Inc, 2002, pp. 94-95

and theoretical models, without violating other theories or causing irreconcilable difficulties in causality. It's not easy to explain even when it works as expected, what to say when it doesn't even follow the rules.

Faulty Theories

The Pioneer spacecraft has shown us anomalous behavior. It was decelerating faster than predicted by the laws of gravity. Other spacecraft, like Voyager, have rockets on board but Pioneer was very simple. It didn't have any rockets to cause problems so it should just fly. But it wasn't flying at the right rate. Nobody knows why.

Discover magazine reported that the U.S. space probes brought this disturbing anomaly to our attention.

> "Pioneer 10, launched in 1972…seems to be defying the laws of gravity. [It] has been slowing down, as if the gravitational pull on it from the sun is growing progressively stronger the farther away it gets."[46]

Other space probes exhibited the same anomalies:

> "Pioneer 10 is not the only spacecraft acting strangely. Pioneer 11, launched in 1973, also slowed down as it pulled away from the sun, right until NASA lost contact with it in 1995. And there's some evidence of similar bizarre effects on two other probes: Ulysses, which has been orbiting the sun for 13 years, and Galileo, which plunged into Jupiter's atmosphere…"[47]

---

[46] "Nailing Down Gravity," *Discover*, October 2003, p. 36
[47] "Nailing Down Gravity," *Discover*, October 2003, p. 36

Fuzzy Math and Faulty Theories

Michael Nieto, is a theoretical physicist at Los Alamos National Laboratory in New Mexico. When asked about these peculiar incidents, he commented:

"We don't know anything. Everything about gravity is mysterious."[48]

His colleague, Thomas Bowles, admitted much the same thing. "Right now, we don't have a theory of how gravity is created."[49]

Three German scientists published a paper confirming that the Pioneer space probe anomalies can't be explained by: additional masses in the solar system, an accelerated sun, the drift of clocks on earth, or cosmic dust.

There are also "flyby" anomalies. These occur when satellites swing by Earth and have a significant unexplained increase of velocity.

The German team also demonstrated that the Astronomical Unit has increased over time. [The distance between the sun and the earth is one Astronomical Unit]. Moreover, comets return several days before their predicted arrival. All of these anomalies were unexplained.[50]

What do we make of these discoveries? Do Newton's laws stop working beyond Earth?

Newton's laws of motion are accurate for things with the mass of an apple or things that weigh a few kilos and travel at a few meters per second. But they break down as we approach the speed of light, or the size of atomic particles. That's why relativity theory and quantum theory became necessary.

We may note that comets and space probes are not subatomic in size, nor approach the speed of light, and yet gravity still did not act as predicted according to our present knowledge. Future discoveries may bring to light additional problems which will necessitate new theories.

---

[48] "Nailing Down Gravity," *Discover*, October 2003, p. 36
[49] Nature Reviews, "Gravity Leaps into Quantum World," January 17, 2002, Tom Clarke, p. 2
[50] Lämmerzahl, Preuss and Dittus, "Is the Physics within the Solar System Really Understood?" Max Planck Institute, April 12, 2006, pp. 1-23

Regarding relativity theory, General Relativity assumes a uniform curvature of space. If indeed, that is the case, then we may question how and why the orbits of planets around celestial bodies are elliptical with varying degrees of eccentricity.

This question is important because a uniform curvature of space should render a uniform degree of eccentricity throughout the cosmos.

The judgment is inescapable: relativity theory cannot fully explain gravity.

9
Quantum Theory

In our courtroom drama we have discovered that the science creation story may be on thin ice. The most common occurrence in nature, gravity, has yet to be fully understood. The vast size of the universe creates problems for our "knowledge" of gravitational force.

Science has left many unanswered questions about gravity on the table. How does the force of gravity hold all bodies in the universe in perfect balance so that they float effortlessly in their orbits even at immense distances? Does quantum theory provide the answer?

Quantum Mechanics

Much of today's confusion in physics is due to Quantum Mechanics. It is often labeled as the spooky realm of quantum theory. Our attorney continues his cross examination in court this morning, focusing on quantum theory and string theory. In terms of making sense of gravity does quantum theory fare any better than Newton's and Einstein's theories?

In their attempt to uncover the physical mechanism of gravity, quantum theorists explain gravitational force as a process of interacting particles. Based on this, they have postulated a particle called the graviton although it has never been detected in experiments. The mass of a graviton is defined as zero and its electric charge is also zero, neither positive nor negative. But it does have spin.

Dr. Paul Davies writes that, "All fundamental particles of matter, the quarks and the leptons possess spin ½. By contrast, the known exchange particles have spin 1 (the graviton, if it exists, would have spin 2)."[51]

Exchange particles are force carrier particles. Forces between particles arise from the exchange of other particles.

By admitting "if it exists", Davies reveals that gravitons are not real world particles like electrons but merely theoretical constructs. Despite the fact that a graviton's existence and properties are only inferred via mathematical theory, physicists still explain that gravitational force "operates by the exchange of gravitons."

If that isn't wispy enough, he has no choice but to admit the following:

> "Nobody has ever detected a graviton directly, because gravity is such a weak force, but its properties can be deduced from what we know about gravitational fields."[52]

Science still doesn't know what gravity is made of, or whether it is inherent within matter. Still, physics courses teach that it's related to the mass of bodies. But if gravitational force is related to sub-atomic gravitons, what is the mechanism for such a weak force-carrier to hold all the stars, galaxies and planets of the entire cosmic creation in its thrall? This weak force governs the entire universe which the small magnet on the refrigerator defies. We are faced with a paradox.

---

[51] Paul Davies, *The Goldilocks Enigma*, Penquin Books, London, 2007, p. 122
[52] Paul Davies, *The Goldilocks Enigma*, Penquin Books, London, 2007, pp. 112-113

Yes, we can see the effects of gravity and can make quite reliable predictions, but that doesn't mean that we fully understand gravity. Newton was able to calculate the effects of gravity but still no one knows if gravity is a wave or a particle.

Do these undiscovered gravitons only carry gravitational force, or is there more to them? Davies enlightens us.

> "Additional information can be encoded in gravitons, which cosmologists believe permeate the universe, although nobody seriously expects to detect any in the foreseeable future."[53]

OK, so now we know this is the belief of cosmologists, not verifiable proof. Davies sums it all up with this revealing information about deducing the concept of gravitons from quantum theory:

> "By using mathematical tricks, physicists are able to dodge round the infinities and still use the theory of quantum electrodynamics to obtain sensible answers to questions about particle masses, energy levels, scattering processes, and so on. The theory remains brilliantly successful. But the fact that infinities occur is a worrying symptom that something is deeply wrong, something that needs fixing."[54]

The attempt to describe the force of gravity by quantum mechanics may also be a dead end street, as Davies candidly points out.

> "...the gravitational force surrounding a particle can be envisaged as a cloud of virtual gravitons. As in the electromagnetic case, infinities follow. But with gravitation there is double

---

[53] Paul Davies, *The Goldilocks Enigma*, Penquin Books, London, 2007, p. 270
[54] Paul Davies, *The Goldilocks Enigma*, Penquin Books, London, 2007, p. 126

> trouble. Any point-like particle (e.g. an electron) would be surrounded by a virtual graviton cloud containing infinite energy. But because energy is a source of gravitation, gravitons themselves contribute to the total gravitational field. So each virtual graviton in the cloud surrounding the central particle possesses its own cloud of yet more gravitons clustering around it and so on *ad infinitum...*"[55]

The conclusion of this theoretical muddle is the following admission.

> "A straightforward quantum description of the gravitational field produces a limitless progression of infinities, ruining any hope of making sensible predictions from the theory."[56]

The technical term for this situation is non-renormalizability.

So the attempt to explain gravity via quantum theory has created problems imposed by recurring infinities. This has confounded the field of particle physics for decades leading to an unavoidable conclusion: gravity is still not understood in terms of the real world.

Moreover, due to baffling equations with recurring infinities, gravity is not even understood mathematically.

The decision is now unequivocal. Quantum Mechanics has not been able to explain the force of gravity. Additionally, quantum theory may nullify one of Einstein's most famous conceptions—space-time. It could indicate the overthrow of Relativity.

It's a little known secret that for sixty years a dark cloud has been hovering over science. Our understanding of the cosmos is based on two separate theories, two sets of laws that do not agree. Relativity and Quantum Mechanics are at war with each

---
[55] Paul Davies, *The Goldilocks Enigma*, Penquin Books, London, 2007, p. 126
[56] Paul Davies, *The Goldilocks Enigma*, Penquin Books, London, 2007, p. 126

Quantum Theory

other, and may not be able to settle their differences amicably.

George Musser states as much in an article published in *Scientific American*, indicating not only a war between quantum theory and relativity, but another one brewing—between relativity and reality itself:

> "After all, relativity is riddled with holes—black holes. It predicts that stars can collapse to infinitesimal points but fails to explain what happens then. Clearly the theory is incomplete... Moreover, quantum theory turns the clock back to a pre-Einsteinian conception of space and time. It says, for example, that an eight-liter bucket can hold eight times as much as a one-liter bucket. That is true in everyday life, but relativity cautions that the eight-liter bucket can ultimately hold only four times as much— that is, the true capacity of buckets goes up in proportion to their surface area rather than their volume. This restriction is known as the holographic limit.
>
> "When the contents of the buckets are dense enough, exceeding the limit triggers a collapse to a black hole. Black holes may thus signal the breakdown not only of relativity but also of quantum theory (not to mention buckets)."[57]

Remember the wild claim from Stephen Hawking on public TV that "the scientific account is complete." In contrast, cosmologists admit privately that the universe is actually a puzzle which they have not yet understood. They are far from a resolution both in theory as well as the practical proof the theory represents.

---

[57] George Musser, "Was Einstein Right?" Scientific American, Sept. 2004, p. 89

So if gravity and even mathematics itself are not rock solid concepts within physics, what does this tell us about the even more mysterious quantum theory? The fluctuating picture of the sub-atomic world does not seem to obey any laws that we know. The law of the quantum world is an oxymoron.

The chaotic nature of quantum theory means nothing is certain. Probability is not a law. It might be a good guess, or it may just be close to the mark. Certainty has gone by the wayside. Today's confusion in physics is due to the spooky nature of quantum theory which leads to conjecture and speculation.

At the end of the 19th century scientists thought they had figured out the basic laws of the universe. Then, Einstein revised our ideas of space, time, and gravity. Quantum Mechanics unveiled the inner workings of the subatomic world, a perspective that was bizarre and uncertain. Is it an actual theory of physics, or just philosophical conjecture about what we do not yet understand?

Each generation of scientists has discovered another unexpected and more puzzling layer of reality, as Fred Hoyle pointed out earlier.

According to Stephen Hawking, "Gravity and quantum theory cause universes to be created spontaneously out of nothing." But we have seen that gravity and quantum theory continue to remain an unsolved puzzle for physicists, so how can he come to such a conclusion? And what does he really mean when he states on TV that the scientific account is complete?

THE UNCERTAINTY PRINCIPLE

In his 2005 book *A Briefer History of Time*, Hawking wrote that the "Uncertainty Principle means that even 'empty' space is filled with pairs of virtual particles and anti-particles."

Virtual particles, like gravitons, are simply theoretical constructs. A particle detector can detect real particles, but virtual particles and gravitons have never been directly observed.

What is the meaning of empty space if it is filled with virtual particles? As far as physicists are concerned, the word "empty" doesn't actually mean empty like you and I understand. The scientific definition of "empty space" is a cosmos filled with particles as Hawking explains:

> "...if 'empty' space were really completely empty—that would mean that all the fields, such as the gravitational and electromagnetic fields, would have to be exactly zero. However, the value of a field and its rate of change with time are like position and velocity of a particle: the Uncertainty Principle implies that the more accurately one knows one of these quantities, the less accurately one can know the other. So if a field in empty space were fixed at exactly zero, then it would have both a precise value (zero) and a precise rate of change (also zero), in violation of that principle. Thus there must be a certain minimum amount of uncertainty, or quantum fluctuations, in the value of the field."[58]

Ironically, Hawking bases his belief on the certainty of the Uncertainty Principle. However, our research has revealed that some very authoritative people in modern physics, like Charles Misner, Kip Thorne, and John Wheeler, have expressed doubts about Heisenberg's principle. Is it possible that the Uncertainty Principle is itself uncertain?

> "The Uncertainty Principle thus deprives one of any way whatsoever to predict, or even to give meaning to, 'the deterministic classical history of

---

[58] Stephen Hawking and L. Mlodinow, *A Briefer History of Time*, New York, Bantam Dell, 2005, pp. 122-123

space evolving in time.' No prediction of spacetime, therefore no meaning for spacetime, is the verdict of the quantum principle. That object which is central to all of classical general relativity, four-dimensional spacetime geometry, simply does not exist, except in a classical approximation."[59]

So take your pick—classical Relativity or the Uncertainty Principle? You can't have both. The popular string theory physicist, Brian Greene, acknowledges the problem that theoretical physics is always in a state of flux, because as one question is put to rest, others arise:

"We then encounter subsequent discoveries that transformed the question once again by redefining the meaning of 'empty' envisioning that space is unavoidably suffused with what are called quantum fields and possibly a diffuse uniform energy called a cosmological constant —modern echoes of the old and discredited notion of a space-filling ether."[60]

Physicists are now faced with an even more vexing problem. Sub-atomic particles appear to have a mind of their own. They don't obey the laws that experimenters expect of them. We saw in the last chapter that the "double-slit" experiment showed how particles behave differently when an observer is or is not taking measurements. Somehow, they "know" how to behave in the different situations.

Physicists do not understand how particles stay in touch with each other, and how this is true even over vast distances. Researchers like Brian Greene have no explanation for this dilemma.

---

[59] *Gravitation*, New York, W. H. Freeman and Co, 1973, 25th printing, pp. 1182-1183
[60] Brian Greene, *The Fabric of the Cosmos*, New York, Alfred A Knopf, 2004, Preface

"At the end of the day, no matter what holistic words one uses or what lack of information one highlights, two widely separated particles, each of which is governed by the randomness of quantum mechanics, somehow stay sufficiently 'in touch' so that whatever one does, the other instantly does too. And that seems to suggest that some kind of faster-than-light something is operating between them. Where do we stand? There is no ironclad universally accepted answer."[61]

The faster-than-light communication between particles infers that light speed may not be the maximum speed in the universe. Moreover, the concept of consciousness pervading the universe via sub-atomic particles creates a further headache for the modern creation story.

THE SPEED OF LIGHT

We have come to know that cosmic inflation exceeded the speed of light, gravitational force might surpass the speed of light, and communication between quantum particles seems to be faster than light speed.

When Einstein proposed that the speed of light was impossible to exceed, the theory of inflation and the communication of quantum particles over vast distances were unknown. The action-at-a-distance nature of gravity was also unresolved.

Based on the three issues above, there may be some doubt whether the speed of light is the maximum speed limit in the universe. How do scientists deal with the anomaly of the speed of light? In Chapter Thirteen we will return to this again in more detail.

---

[61] Brian Greene, The Fabric of the Cosmos: Space, Time and the Texture of Reality, New York: Alfred A. Knopf, 2004, pp. 117-118

For now, Nobel Laureate Steven Weinberg has admitted: "The techniques by which we decide on the acceptance of physical theories are extremely subjective."[62]

Similarly, Stanislaw Ulam clarifies the issue for the general public.

> "I should add here for the benefit of the reader who is not a professional physicist that the last thirty years or so have been a period of kaleidoscopically changing explanations of the increasingly strange world of elementary particles and of fields of force. A number of extremely talented theorists vie with each other in learned and clever attempts to explain and order the constant flow of experimental results which, or so it seems to me, almost perversely cast doubts about the just completed theoretical formulations."[63]

When it comes to establishing the implications of quantum theory, Stephen Jay Gould, a leading figure of 20th century physics, is not convinced that we have really understood the laws of the cosmos.

> "It's probably because we're not thinking about them right. Infinity is a paradox within Cartesian space, right? When I was eight or nine I used to say, 'Well, there's a brick wall out there.' Well, what's beyond the brick wall? But that's Cartesian space, and even if space is curved you still can't help thinking what's beyond the curve, even if that's not the right way of thinking about it. Maybe all of that's just wrong! Maybe it's a universe of fractal expan-

---
[62] John Horgan, *The End of Science*, New York, Broadway Books, 1996, p.74
[63] Stanislaw Ulam, *Adventures of a Mathematician*, University of California Press, 1976, p. 261

sions! I don't know what it is. Maybe there are ways in which this universe is structured we just can't think about."[64]

In a *Discover* magazine interview, Lewis Thomas admitted much the same thing as Gould, but he took it to an even deeper dimension.

"The principle discoveries in this century, taking all in all, are the glimpses of the depth of our ignorance about nature. Things that used to seem clear and rational, matters of absolute certainty—Newtonian mechanics for example—have slipped through our fingers, and we are left with a new set of gigantic puzzles, cosmic uncertainties, ambiguities. Some of the laws of physics require footnotes every few years, some are canceled outright, some undergo revised versions of legislative intent like acts of Congress."[65]

Today there are so many observations in nature, along with an over-abundance of data that scientists are at a loss to make sense of it all. This is especially true in the field of cosmology.

We collect an assortment of facts and figures and attempt to offer an interpretation. But correct interpretation of data is the key to knowledge, and most scientists simply don't have the insights. The universe is beyond their simple understanding as well as beyond their complex understanding.

Astronomer Robert Dicke openly declared that he is puzzled by what he sees in the universe.

"There are peculiar puzzles about this Universe of ours. As it gets older, more and more of the

---

[64] Interview with John Horgan, cited in *The End of Science*, New York: Broadway Books, 1996, p. 125
[65] Lewis Thomas, "Making Science Work," Discover, March 1981, p. 88

> Universe comes into view, but when new matter appears it is isotropically [evenly] distributed about us, and it has the appropriate density and velocity to be part of a uniform Universe. How did this uniformity come about if the first communication of the various parts of the Universe with each other first occurred long after the start of the expansion?
>
> "The puzzle here is the following: how did the initial explosion [the Big Bang] become started with such precision, the outward radial motion became so finely adjusted as to enable the various parts of the Universe to fly apart while continuously slowing in the rate of expansion. There seems to be no fundamental theoretical reason for such a fine balance."[66]

Of course, this admission is just the tip of the iceberg. By the end of the 20th century it became clear that cosmology was baffled. Some cosmologists are now beginning to think they're on the wrong track.

John Horgan declares in his book *The End of Science* that many scientists are no longer completely open and straightforward.

> "...sometimes the clearest science writing is the most dishonest... Much of modern cosmology, particularly those aspects inspired by unified theories of particle physics and other esoteric ideas, is preposterous. Or, rather, it is ironic science, science that is not experimentally testable or resolvable even in principle and therefore is not science in the strict sense at all. Its

---

[66] Robert H. Dicke, "Gravitation and the Universe, Jayne Lectures for 1969," *American Philosophical Society*, Philadelphia, 1970, p. 61-62

primary function is to keep us awestruck before the mystery of the cosmos."[67]

This brings us to the next contender to try and explain gravity—string theory. It's a recent scheme of physics that is exciting physicists.

String theorists say that particles are not what science has taught. If we look up close, particles are really tiny vibrating string-like objects. *Oops, looks like we made a mistake, but we're on the right track now.*

STRING THEORY

In the 1980s, cosmologists dreamed up a lyrical sounding idea—string theory. String theory starts with the concept of a string that vibrates, then doing physics with that. What we previously misconstrued to be particles are now described as loops of string so infinitesimally tiny that we mistook them for point particles.

Suddenly, strings were the basis of everything in existence. The new theory promised to solve the mysteries of the universe. The result? Well, the entire conception of particles was redefined. Paul Davies reveals the nature of strings:

> "In the simplest version of the theory the strings form closed loops, but they are so tiny that it would take a chain of a hundred billion billion of them to stretch across a single atomic nucleus."[68]

Point particles are actually only vibrations, or oscillations, of the strings. It's like using guitar, bass, and violin to make different notes from vibrations of strings in different registers.

---

[67] John Horgan, The End of Science: Facing the Limits of Knowledge in the Twilight of the Scientific Age, New York: Broadway Books, 1997, pp. 93-94
[68] Paul Davies, *The Goldilocks Enigma*, Penquin Books, London, 2007, p. 127

The vibrating strings produce music.

What makes string theory popular is that now physicists have a model of vibrating strings which comprise the other sub-atomic particles, the quarks and leptons of atoms. It means that strings are considered fundamental.

The theory posits that the vibrating strings are pliable, compared to the rigidness of particles. Moreover, the amount of vibration possessed by a string defines its size and shape, which together determine its function.

When physicists explored the mathematics of string theory in detail, they made another discovery. It wasn't just strings that were involved in the physics; there were membranes that also vibrated. So string theory evolved into M-theory. It was like adding drums to the guitar, bass, and violin to form a band to broaden the scope of the music.

A membrane was considered to be like a sheet. If the sheet was rolled up like a drinking straw it would look like a string. So physicists began saying that the basis of everything in nature consists of vibrating membranes and vibrating strings.

Since the new theory postulates that everything in existence is built of these strings, the first question is what are the strings made of? No clear decision is forthcoming so don't expect a response any time soon.

Can string theory finally give us the answers to explain gravitational force? That's a far-fetched hope, although strings are thought to be some primordial fundamental entity. In the last three decades physicists have moved no closer to understanding what this might be since the nature of the strings is so elusive.

Other problems exist. The strings are so small there is zero hope of detecting them with modern technology. We can't even confirm their existence with clear predictions that might be sensed by today's instruments.

It soon became apparent that string theory was not just one theory. Rather, it produced five theories that were basically different from each other. There were theories with closed strings, with open strings, and theories that required different dimensions to function properly. It looked like the multiplicity of different possibilities was just another muddle that couldn't be clarified.

Before the entire enterprise was derailed, however, a group of theoretical physicists suggested that the strings moved in ten dimensions instead of three. In order to make sense of the mathematics, physicists had to add six extra dimensions of space.

But where are these extra dimensions?

It became necessary to introduce a mechanism that hid them from our view. They called that idea compactification. String theory now had the three dimensions we all live in, six compactified theoretical dimensions that are unobservable, plus time. Paul Davies comments:

> "String Theory deals with a rarefied domain of ultra-high energies and ultra-small distances, and doesn't so far have much to say about the real physics that takes place in the laboratory."[69]

Although physicists began to accept that particles in the universe consist of vibrating strings, there was still the problem of five different string theories. Edward Witten was the brilliant quantum mathematician who found the solution to unify the five different versions of string theory. He called his new perspective M-theory. It's purported to be the mother of all theories.

Will M-theory finally reveal the answer to what causes gravitational force?

---

[69] Paul Davies, *The Goldilocks Enigma*, Penquin Books, London, 2007, pp. 127-128

M-THEORY

Ed Witten proposed that the five string theories we discussed earlier was just a different way of looking at the same thing. An appropriate analogy might be the story of the five blind men who latched onto five different parts of an elephant. Each one gave a different version of how they described the beast.

The five different string theories were all contenders for the long-hoped for Grand Unified Theory, but they were found not to be in harmony with each other. It was another embarrassing situation for physicists. In hopes of a resolution they began fixing theories and projecting other theories because they believed the mathematics. Each theory appeared equally valid, but did the equations give a true picture of reality?

We already learned from Paul Davies that the equations that govern the full M-theory have yet to be solved. "In spite of this murkiness," he assured us, "M-theory has generated tremendous enthusiasm."[70]

We know that Stephen Hawking is one of the enthusiastic fans. He has staked his entire reputation on a grand design scheme based on M-theory.

While not sharing Hawking's fervent zeal, Davies does acknowledge that M-theory represents progress even though the mathematical fragments are obscure and abstract. Why? Because they "provide a tantalizing glimpse of a still unexplored theory of extraordinary power and elegance that may yet turn out to be the key to the universe."[71]

But first, string theorists had to resolve problems with ten dimensional reality. By adding an 11th dimension Witten unified some string theory problems and the new perspective became known as M-theory.

An explanation was proffered that strings moving in ten

---

[70] Paul Davies, *The Goldilocks Enigma*, Penquin Books, London, 2007, p. 129
[71] Paul Davies, *The Goldilocks Enigma*, Penquin Books, London, 2007, p. 130

dimensions might actually be like sheets moving in eleven dimensions. When the sheet was rolled up like a drinking straw it looked like a string. The sheets were called membranes, or 'branes' for short.

M-theory purports to explain how the tiniest as well as the largest things in the cosmos work. Its ultimate achievement is the recognition that to make sense of things we need to assume the universe exists in eleven dimensions. The downside is these dimensions are unseen and undetectable.

If this explanation isn't weird enough, M-theory proposes that six or seven dimensions are tiny and right in front of us. But we can't sense them. The theory also proposes that we live on a giant energetic membrane. Our universe is tethered to this 'brane' by extra dimensions that are invisible.

There are branes so close to each other that these giant walls of energetic matter float side-by-side like gigantic sheets in a bigger structure called 'the bulk' or hyperspace. A universe may be attached to one brane, or a universe might occupy an entire brane.

So our real-world observable universe may be sitting on the surface of a giant brane which is drifting within the cosmic ocean of space. Think of being on a cruise ship. It's a crude example but above the water is one dimension, a brane, and below the ship is another dimension, the realm of aquatics. We have no connection to them and they have no connection to us because we're both in different dimensions.

To accept that invisible extra dimensions connect us like an umbilical cord to a hyperspace where giant membranes exist, requires a huge leap of faith. These extra dimensions and the branes we are supposedly attached to cannot be detected with today's technology. In spite of this, physicists have high hopes the theory will be verified in the near future. In other words, it's more faith than science at present.

GRAVITY AND M-THEORY

If M-theory can at least explain gravitational force then we have cause for optimism. Until then, it's only speculation. Scientists at Harvard University continue their work to unravel the mysteries of gravity. Lisa Randall gives us an update.

"It turns out there are very new ideas on how to explain the weakness of gravity if we have extra dimensions." She believes that gravity can be better understood if there are extra dimensions in the universe.

When M-theory was proposed as a solution to the mystery of gravity, Randall and her colleagues thought it could provide the explanation they were seeking. She wondered if gravity only appeared to be weak, but fundamentally might be as strong as the other forces. Could its strength be diluted through the extra dimensions that weren't detectable?

Randall attempted to calculate how gravity could leak from our membrane universe into "empty space." But she couldn't get the mathematics to work. Since M-theory posits that there are eleven dimensions, was gravity leaking from the cosmos into some empty space of the 11th dimension?

When she heard that there could be another brane in the 11th dimension, she had a curious thought. If gravity wasn't leaking out of our universe, might it be leaking into it? On that other brane gravity could be stronger, but was only a faint signal when it reached us. What if it came from another universe?

She reworked her calculations and everything fit perfectly. If the math fits, then it works. Like magic! Mathematics saved the day, once again.

"If you were to imagine that there were two membranes," Randall explains, "say there's one in which we sit, and one in which, if there's other stuff it sits there, but not our particles, not the stuff that we're made of, and not the stuff that we see forces associated with. If we lived anywhere else in the extra dimension we would see gravity as very weak because it's mostly spending its time near the other brane. We only see the tail end of gravity."

Randall's idea opened up Pandora's Box, and physicists jumped into the 11th dimension in an effort to resolve intractable problems. Taking refuge of mathematics, they found a perfect explanation was always due to another parallel universe. Everywhere they looked they began to find all sorts of parallel universes they never knew existed.

The weakness of gravitational force may now have an explanation, but only by introducing the idea of membranes and theoretical dimensions beyond our ability to test or observe. The jury is asked to consider whether scientists have finally come to grips with gravitational force or are still struggling for a solution.

10

Parallel Universes

Everyone in court this morning is buzzing about yesterday's interrogation. Today, our attorney has promised to extract an in-depth view of parallel universes, what various physicists think about the multiverse idea, and what direction it's taking cosmology.

This segment might sound like science fiction. Yet many 21st century cosmologists seriously accept the conclusions that we will introduce. These conclusions are derived from the mathematics that many physicists earnestly believe represents reality.

THE MULTIVERSE

The multiverse concept was broached by Stephen Hawking in *The Grand Design* but now we're going to have a good long look at it.

We already learned that M-theory posits 10^{500} possible universes, collectively called the multiverse. In our observable universe there are about 10^{80} atoms. Thus, the number of parallel worlds according to M-theory is unimaginably greater than the total number of atoms in the entire cosmos.[72]

Of course, physicists are not talking about the real world we can observe. The multiverse definition posits an unlimited number of unobservable and undetectable worlds outside our universe. It's comparable to the heaven and hell dimensions within religion which are also unobservable, undetectable, and outside our universe.

We should establish first off, that the parallel universes idea is not a theory. Rather, the idea grew up in predictions from other theories. These unseen worlds are based entirely on mathematics, which we now know can be an unreliable "proof" of reality.

There is not a shred of empirical evidence to support the hypothesis that parallel universes are real. This begs the question: How can physicists speak about undetected realms that are simply mathematical ideas devoid of empirical data that are akin to religious belief? This belief in a multiverse might be the textbook example of what Richard Dawkins calls "blind faith without evidence."

Many physicists decry the modern addiction to mathematical equations for constructing models of reality, and from that, basing conclusions about the nature of the real world. Previously, if scientists couldn't observe phenomena, say religious miracles, they excluded them from scientific consideration.

With M-theory, now physicists routinely include unobservable phenomena for scientific deliberation because they can manipulate the numbers to make things look plausible. It means that anything and everything could be possible in some parallel universe, even if undetectable and untestable.

---

[72] Paul Davies, *The Goldilocks Enigma*, Penquin Books, London, 2007, p. 192

Parallel Universes

In 2008, the History Channel aired a show called *Parallel Universes*. Our attorney sets up his projector again to view the contents of that program. He notes that a TV show is basically for entertainment and should be understood in this light. But what about the physicists who are interviewed? Do their views represent scientific knowledge or entertainment value?

After a sensationalist introduction a narrator speaks the following: "Imagine another world; a whole other universe with a solar system and a planet exactly like ours. On this parallel earth there lives an exact copy of you. You could be leading exactly the same life but in another universe."

"Now, imagine an entirely separate universe where you could be living a slightly different life at the same time as you live this life. It may seem fantastic, incredible, and impossible, but there could be many other universes out there. And now some of the world's leading physicists believe that they have evidence to prove it."

Unfortunately, the "evidence to prove it" is never presented. What *is* presented are hypotheses without data. After exhaustive research it was impossible to find any factual data to substantiate the parallel universes conception. But lack of evidence doesn't seem to stop modern cosmologists.

Michio Kaku, author of the book *Physics of the Impossible* and physics professor at City College of New York, explains the present state of the modern physicist's universe. "We are experiencing an existential shock. Our world view has been shattered with the realization that, *Yes, there could be parallel universes.*"

Of course, "could be" means this is a belief system. The hypothesis is highly speculative. It's not the familiar ironclad science that we're accustomed to.

Dr. Kaku continues: "Revolutionary developments have

changed the entire landscape. Data from outer space has given us a new look at cosmology. And satellite data indicates there could be parallel universes."

The term "data from outer space" is misleading. The reference here is to experiments that are being conducted to determine if the universe is infinite. An infinite cosmos could allow the possibility of unseen parallel universes to exist. Of course, such experiments don't prove they exist, just that they could. This is the data being referred to by Kaku.

Einstein established that matter curves space in predictable ways. It means that if the universe has a different density of matter it should have a different shape. Einstein's theories allow for three possible scenarios: negative curvature, in which the universe looks like a saddle; positive curvature, where the universe is a hypersphere; and flat, where the overall density of matter doesn't warp space. In a flat universe, photons traveling in a constant direction could never return to a starting point.

We should note that "flat" does not mean two dimensional; rather it means a flat three dimensional space that is not warped. The density of matter is denoted by the symbol *omega*. For the universe to be flat, the total matter must reach critical density where *omega* equals one. In a saddle-shaped universe, *omega* is less than one; in a spherical universe, *omega* is greater than one.

Astronomers determine the value of *omega* by measuring the way space bends beams of light. The light they measure is the cosmic microwave radiation glowing at the farthest reaches of the cosmos. The distortion in the microwave signal reveals the shape of the intervening space.

In a flat universe, patches of background radiation would be close to their predicted size. A spherical-shaped universe would magnify the patches of background radiation. While in a saddle-shaped universe, distinct patches of the microwave background would look smaller than the predictions.

When density is equal to 1 degree, the topology of the universe will be flat like a table top. If less than critical density we get negative curvature akin to the surface of a saddle. Therefore, the topology of space would be open and infinite. If critical density shows positive curvature the universe will curve back in on itself to form a hypersphere, making it finite in size and volume.

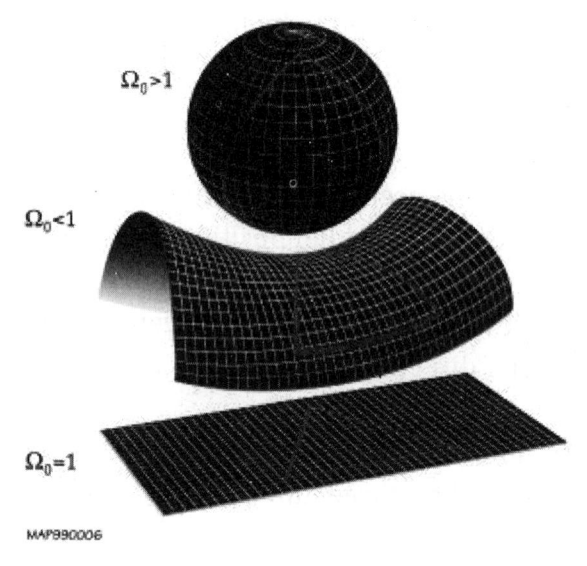

Credit: Wikipedia.org

In the diagram above we can see the possible topography according to omega. Again, for parallel universes to exist the cosmos must be infinite.

Cosmologists can also determine the shape of the universe by using the geometry of triangles—triangulation. As luck would have it, the Cosmic Microwave Background provided an extraordinary opportunity for a triangulation experiment.

The experiment involved shooting laser beams out into deep space to form a giant light triangle with the two farthest points of the CMB. If the universe was flat the angles should add up to exactly 180 degrees, even with an enormous triangle.

More than 180 degrees would indicate positive curvature while less than 180 degrees would imply negative curvature.

The WMAP satellite was utilized to measure the geometry of the universe. When the measurements were in, the angles added up to exactly 180 degrees. Therefore, this experiment suggested a flat and infinite universe as depicted in the diagram below.

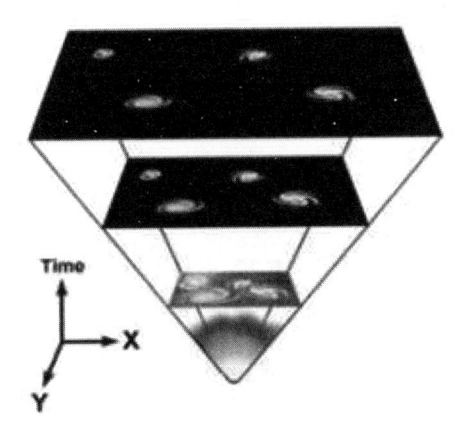

(Flat Universe: Creative Commons Attribution-Share Alike 3.0 Unported | Bjarmason)

At this point in the discussion it's vital to distinguish between "universe" and "observable universe." We note that the "observable universe" represents all that exists within which light has had time to reach us. On the other hand "universe" constitutes everything in existence, what we can and cannot see or detect.

There is no way to acknowledge anything beyond our observable horizon. So, all possible observations are confined to the observable universe. Many people use the term "universe" when they actually mean the "observable universe."

The "satellite data" mentioned earlier by Professor Kaku refers to data from the Microwave Anisotropy Probe, or the WMAP satellite, which measured the radiation left over from the Big Bang. In other words, WMAP was originally designed to record the very earliest signs of creation.

Some cosmologists believe that the remarkable satellite images reveal the true shape of the universe; that it's flat. What they mean is that measurements determine that omega = 1.

When they conclude that "WMAP provides data supporting a flat universe", they should really be saying that "WMAP provides data that the observable universe appears to be flat."

The extraordinary image below reveals the status of the universe shortly after the Big Bang. The color differences are temperature fluctuations.

Credit: NASA/WMAP Science Team

Max Tegmark of M.I.T. explains what the image tells us. "What we have here are baby pictures of the universe, what it looked like when it was only 400,000 years old. We're looking so far back in time that the galaxies hadn't even formed yet. We just had this dead, diffuse gas which gradually, over time, clumped into galaxies, stars, planets."

Tegmark's comment about "dead, diffuse gas" indicates his view that it's devoid of consciousness. So how did consciousness evolve from dead diffuse gas? No answer has ever been offered.

Anyway, the WMAP data is challenging what scientists thought they knew about the universe. Every 20th century physicist assumed the universe was curved, as Einstein had posited. But in the 21st century, this is no longer the case. With the same degree of certainty, cosmologists in 2013 accept the universe to be flat within a 0.4 percent margin of error. This suggests the cosmos should be infinite.

Another perspective is that the universe may have inflated so quickly and so enormously that it only looks flat, like a gigantic balloon. Thus, even if the cosmos has curvature, the observable universe would still appear flat due to the limitations of scientific instruments. So at present, there really is no guarantee or proof of its infinitude.

Several theories are accepted by scientists to bolster the hypothesis that the cosmos might be infinite. The caveat, of course, is the universe had a beginning. Science has no way of observing beyond the cosmological horizon, so we lack information about events that happen beyond that limited horizon. One thing is certain—the observable universe has finite extension which is far greater than any of our instruments can detect.

According to inflationary theory, the universe expanded faster than the speed of light. But if it was finite speed, could it yield infinite size over a finite period? The real challenge comes from the rate of universal expansion being a finite figure. So the idea of an infinite universe due to cosmic inflation from a Big Bang is mathematically impossible.

Another way to determine whether the universe is infinite is via mathematics. This so-called "data" allows for the *possibility* of parallel universes. The equations could work if there were an infinite number of such parallel universes, but such mathematical evidence is unverifiable.

Nevertheless, many physicists are excited at the prospect. "New theories, called string theories, are giving us worlds of higher dimensions," says Professor Kaku. "Quantum physics at the microscopic scale is also revealing to us the fact that there could be parallel universes."

His proposal comes from the spooky world of quantum mechanics. Electrons may be at multiple places at the same time. Werner Heisenberg's Uncertainty Principle has enshrined this amazing quantum phenomenon.

"As crazy as it sounds," Max Tegmark comments, "not only does quantum physics tell you that a little particle can be in two places at once, but the so-called Heisenberg Uncertainty Principle tells you that sometimes the particle even must be in two places at once."

As proof of this, scientists use a laser light and a glass apple to show how light particles, or photons, appear in several places at the same time.

Tegmark presents his glass apple to demonstrate how it works: "This shows that photons, the little particles of light coming out of my laser, can end up in several places at once. And since we are made of little particles, that means if they can be at several places at once, so can we."

His argument sounds fallacious. It's like saying that since humans have the same molecules as fish, therefore we can also live in the sea.

From a physics perspective, the micro-world and the macro-world work differently. There is no analogous connection to make them behave the same way. Factually, quantum theory only applies to sub-atomic particles. And now it should apply to macro objects like humans in parallel universes?

We have already learned that several prominent physicists have cast doubt on the Uncertainty Principle, while others hold it as inviolable. From observations that electrons and photons appear in different places simultaneously, and the singularity idea from which the Big Bang emerged, Michio Kaku promotes the following conclusion.

"The universe at one point was actually smaller than an electron. If that's true and if electrons are described as being in many places at the same time in parallel states, this means that the universe also exists in parallel states and you inevitably get parallel universes."

We know that 'if' is a small word with a huge meaning. The fallacy of Kaku's argument is that he equates the actions of sub-atomic particles to infinite universes. Does an infinite universe, even infinitely compacted in size, behave like a sub-atomic particle?

Like the Tom and Jerry cartoon characters, Quantum theory and Relativity theory are totally incompatible and constantly at war with each other. The way matter behaves at the microscopic level contradicts how it acts at the macroscopic level. Thus, another great enigma of modern physics is that our universe is based on two sets of laws that don't agree. Resolving this contradiction has eluded everyone.

In his 2009 *Discover* interview Roger Penrose points out a further problem with the many worlds interpretation.

> "What can you do with it? Nothing. You want a physical theory that describes the world that we see around us. That's what physics has always been: explain what the world that we see does, and why or how it does it. Many worlds quantum mechanics doesn't do that. Either you accept it and try to make sense of it, which is what a lot of people do, or, like me, you say no—that's beyond the limits of what quantum mechanics can tell us…"

> "My own view is that quantum mechanics is not exactly right, and I think there's a lot of evidence for that. It's just not direct experimental evidence within the scope of current experiments."

Modern ideas in theoretical physics are becoming increasingly surreal. Penrose lays the blame for this on quantum theory, even though quantum mechanics has a lot of experimental support.

Are we finally on the verge of a breakthrough? The proposed solution is M-theory, which only works in eleven dimensions. Is this really a theory of physics or a new philosophy? And if M-theory fails to provide a testable solution, on what basis should we believe it?

Along with other physicists, Kaku promotes the parallel universe idea. "We used to say universe—'uni' meaning one. A one world theory. Everything there is, everything we can see, is the universe. But now we have a multiverse idea where there are unseen worlds. Worlds that we cannot see, worlds that we cannot touch."

This sounds like the traditional creation story. God also exists in a realm that can neither be seen nor touched. That idea was originally declared by science to be superstition; it lacked credibility because it couldn't be verified by data. But now physicists are taking a page from the theologian's book.

Nonetheless, the parallel universe hypothesis proposes numerous discrete scenarios. Max Tegmark and Brian Greene use classification formats to rank the various theoretical types of parallel universe that could comprise a collective multiverse.

Tegmark ranks according to four levels, and Greene according to nine types. These projections are explained via M-theory. Keep in mind that M-theory consists of five different string theories and eleven dimensions of space. The implications are indeed staggering.

Level One Parallel Universe

A Level One parallel universe is beyond our cosmological horizon (beyond what our instruments can detect) although still an extension of our own cosmos. The Level One kind is based on the assumption that the cosmos is infinite and contains an infinite number of smaller parallel universes (like a patchwork quilt) all having the same physical laws and physical constants. Apparently, these "smaller universes" must be finite otherwise how could they be accurately described as smaller.

Mathematical probability suggests that there is an exact copy in infinite space of the solar system, the earth, and all of us. What does it mean? Michio Kaku offers a clue.

"It means that in one universe, Elvis Presley is still alive." Well then, it could mean that John Lennon is also alive in a parallel universe. In another, Mozart never wrote any symphonies. How about in one of the parallel universes the Nazis won World War Two? In another, that war never happened. The list goes on...

Dr. Aleksey Fillipenko describes Level One parallel universes this way:

"Those universes are so far away that the light hasn't reached us. That's the kind of universe which is really sort of part of our giant space, but it's so far away that we can't see it. We think that it was born in a very very small state and then there was a kind of weird energy that pushed it apart by a tremendous amount. It just went WHOOOSH!"

Fillipenko is referring to cosmic inflation which accounts for how our universe suddenly and massively grew after its inception. The theory of inflation proposes that our universe is infinite, which means the rate of inflation would have to be infinite. The question remains: if the rate of inflation was finite, how could the universe become infinite?

Regardless, Max Tegmark buys the inflation idea. "The best theory we have right now for what made space so big, is the theory of inflation. It says, in fact, that space isn't just big or huge, that it's really infinite. It goes on forever, which means there aren't just one, two, three, but infinitely many other regions to space just as big as our universe."

Fillipenko believes that cosmic inflation produced the parallel universes. "So there's just this enormous number of other spaces out there that we can't see. We actually think that there's a gigantic number, perhaps even an infinite number of Level One universes. Then there's a copy of me and you and everyone else out there somewhere."

The old scientific concept was a one-world theory. Everything observable or detectable we called the universe. But radical new concepts are leading to an entirely different definition of the cosmos.

These physicists maintain that the quantum principle which creates many versions of subatomic particles, also creates copies of each person, and also creates entire universes. Thus, if there are an infinite number of other universes, and other earth-type planets, and infinite copies of every one of us, then for each life every possible outcome has to occur.

This means that for the new multiverse idea anything that is physically possible does actually occur in some parallel universe.

Aleksey Fillipenko supports this concept. "If the universe is infinite and there really are all these Level One universes in the infinite multiverse, all those other possibilities did, somewhere, happen."

Of course 'if' means conjecture, not fact. But according to expert opinion even more mind-boggling kinds of parallel universes exist. If the Level One parallel universe doesn't quite make sense to you, then you may not be ready for the Level Two kind.

Level Two Parallel Universe

A Level Two parallel universe is made up of colossal cosmic soap-like bubbles which float in hyperspace, and all with different physical constants. Within every one of these independent bubbles is an entire universe. The premise of the Level Two concept is that at the moment of creation our universe radically and instantly inflated into a massive mega-bubble.

Our specific universe is just one mega-bubble in a cosmic crowd of super bubbles which are floating in a foaming sea of other super bubbles. These cosmic bubbles clash with each other thereby spawning new universes. All together, these super bubbles form the Level Two parallel universe. And within it are an infinite number of Level One parallel universes.

To clarify this situation, Tegmark comments:

> "The Level One and Level Two universes are all in our same one space. There's only one space. Then there are different regions that we call Level One and Level Two parallel universes. What we call the Level Two multiverse is really best thought of as a tree or a fractal structure, where you have a region of space expanding like crazy and sprouting off other regions which then expand and sprout off other regions."

Some physicists are cautious about claims that various experiments determine the flatness of the universe. They prefer the term "almost flat" because a gigantic cosmic bubble may appear flat to our instruments. Michio Kaku considers the universe in this way.

> "I tend to think that the universe is, in fact, a soap bubble of some sort, but it is bent so slightly that we can't see it. In this new paradigm, soap bubbles can form, reform, they can split. It's dynamic—universes being created out of nothing, universes butting off other universes; a multiverse of universes, each one popping into existence, popping out of existence, perhaps colliding with each other."

The scenario is a chaotic, eternally branching set of universes colliding and spawning off from their predecessors. And this infinite set of universes is the multiverse. The technical term for this world-shattering process is Bubble Nucleation. Max Tegmark explains the term for a general audience.

> "Bubble Nucleation is the geek speech phrase for the process where you have this inflating strange material, and then a little piece of it stops inflating, and causes a bubble shaped region expanding around it to also stop inflating. And then you create in this bubble a nice calm region of space where it will eventually form galaxies, stars, planets, and even people like us. So we are the children of the bubble."

Tegmark obviously has a sense of humor. In the traditional creation story we are the children of God.

To clarify his explanation: the multiverse continues inflating forever but some regions of space cease inflating to form distinct bubbles, akin to gas pockets in a loaf of rising bread.

If the Level Two parallel universe idea is established, the nature of the cosmos is more astonishing than science fiction writers ever imagined.

LEVEL THREE PARALLEL UNIVERSE

The Level Three concept invokes the many-worlds interpretation of quantum mechanics. This idea reveals an even more fantastic explanation of parallel universes that exist in Hilbert space which can have infinite spatial dimensions. In both the Level One and Level Two kind we learned that there are replica universes separated from us in time and space.

In the Level Three parallel universe these copies of ourselves are living right here in exactly the same space and time. We don't perceive them because they are in a different dimension than our space. To boggle your mind even further, there are an infinite number of them.

This controversial many-worlds conception has astonishing consequences. In a Level Three parallel universe a tiny quantum difference in thought can change your entire world. The instant you have a fleeting thought, your body makes a quantum leap into a parallel universe in a completely different dimension.

Max Tegmark offers this explanation. "If I'm walking on the pavement and make a snap decision whether I'm going to go left or right, if my decision depends on what some little particle in my brain was doing, then I will actually end up doing both. And my life effectively splits into two parallel realities." No wonder he is known as Mad Max.

Can a fleeting thought send people into an extra dimension, and on to a parallel universe? Joe Lykken, a physicist at the Fermilab particle collider in Illinois, explains this phenomenon.

"It's like the butterfly that flaps its wing and makes the hurricane. Tiny microscopic events would change the course of history. So every historical event actually happened in every possible way in some branch of this ever splitting many worlds universe."

So in one parallel reality, England defeated the American settlers and the United States is still a colony of the British? In another reality, there were no settlers because Europeans never colonized the Americas. Like endless roulette wheels where all possible numbers come up, mathematical probability promises that all possible universes and outcomes will occur in a multiverse.

Even the impossible might become probable in some parallel universe. You may be watching television peacefully in your home while other invisible worlds are raging all around you.

Professor Michio Kaku offers this fantastic scenario.

"These parallel universes are in your living room. This means that in your living room there are dinosaurs. You can't hear them, you can't see the dinosaurs rampaging throughout your living room, but they're there."

Is this what he teaches his physics students? Is there any evidence to back up his assertion?

LEVEL FOUR PARALLEL UNIVERSE

Max Tegmark shocked the world when he claimed there could be one other kind of parallel universe. This level considers any and all universes that can be described by different mathematical structures to be equally real.

Tegmark assumes a reality exists which is independent of us. He says most physicists agree with this concept. However, it's comparable to the explanation of God's realm which is also an independent reality. Religion has taught this for millennia. Science originally labeled it a bronze-age superstition, but now physicists are utilizing a similar idea by positing these mathematical structures.

The Level Four varieties are created either by quantum fluctuations, or by membranes clashing. The creation that results is radically different. In a Level Four parallel universe all the rules fly out the window, comparable again to supernatural ideas. If the laws of physics are completely different in a Level Four parallel universe, then they could be supernatural unless Tegmark has data to substantiate that they are definitely not supernatural.

In any event, the physics describing these universes is totally different to the math and physics we know, so that galaxies, stars, and planets might not have formed and life as we know it would not exist.

Parallel Universes

Brian Greene, as his contribution to M-theory, suggests there are 9 types of possible multiverse configurations containing parallel universes. These are listed in Wikipedia:

1. The quilted multiverse: With an infinite amount of space, every possible event will occur an infinite number of times. However, the speed of light prevents us from being aware of these other identical areas.

2. The inflationary multiverse: This type is composed of various pockets where inflation fields collapse and form new universes.

3. The brane multiverse: This type follows from M-theory and states that each universe is a 3-dimensional brane that exists with many others. Particles are bound to their respective branes, except for gravity.

4. The cyclic multiverse: This type has multiple branes (each one a universe) that collided, causing Big Bangs. The universes bounce back and pass through time, until they are pulled back together and again collide, destroying the old contents and creating them anew.

5. The landscape multiverse: This type relies on string theory's Calabi–Yau shapes. Quantum fluctuations drop the shapes to a lower energy level, creating a pocket with a different set of laws from the surrounding space.

6. The quantum multiverse: This type creates a new universe when a diversion in events occurs, as in the many-worlds interpretation of quantum mechanics.

7. The holographic multiverse: This type is derived from the theory that the surface area of a space can simulate the volume of the region.

8. The simulated multiverse: This type exists on complex computer systems that simulate entire universes.

9. The ultimate multiverse: This type contains every mathematically possible universe under different laws of physics.

These classifications of undetectable universes beg the question how reputable physicists can have a strong belief in something undetectable and unverifiable. Is this similar to a religious belief? Science is about evidence; it's not about belief. And since science has yet to deliver evidence, it is acceptable not to believe the multiverse idea.

Because there is no evidence, Tegmark bases his multiverse belief via *modus ponens* logic: that if X implies Y and X is true, then Y must also be true. He argues that if theory X has enough experimental support to be taken seriously, then we also have to take its predictions Y seriously even if the predictions are untestable.

In other words, several theories imply that diverse types of parallel universes could exist. According to *modus ponens*, if we take those theories seriously, we have to take parallel universes seriously.

His entire argument hinges on the meaning of "if X implies Y". The point is: with what level of certainty does X imply Y? There are strong implications and loose implications. If I'm a messy person it implies my room is messy. But with what degree of certainty? Have I never had a tidy room?

If the truth of X necessitates the truth of Y, then if X is true, Y is also true. It's not enough to have a loose relationship, say in 80 percent of cases.

For example, Tegmark is a fan of Richard Feynman who was against blind following, thus Mad Max is not a blind follower of Feynman. But there might be fans of Feynman who didn't study his work seriously and may be blind followers. In this case X implies Y fails. So it really depends on the certainty of the relationship between X and Y.

To bolster his argument, Tegmark says we take General Relativity seriously because of the strong predictability quotient for observable events. Therefore, we must accept predictions of GR for what happens inside a black hole. For a theory to be testable (and hence scientific), he assures us we don't have to be able to test all its predictions, merely one of its predictions.

But does the physics break down in extreme circumstances approaching the infinite mass inside a black hole? We have only seen what GR predicts outside a black hole. Tegmark needs to offer evidence that it works the same in extreme circumstances and can predict results with the same reliability.

For less proven theories, the doubt factor is even greater. Therefore, just because X is true and implies that prediction Y is true, it does not imply Y is true if unobservable factors or infinite mass affect Y. Physicists must ascertain if the truth of X results in the truth of Y in every case, not just in one case as Tegmark claims.

If E is evidence for hypothesis X, and X implies Y, then E is evidence for Y, seems to be the entire basis for Tegmark's multiverse belief. Yet the *modus ponens* argument is not supported unless there is 100 percent certainty of the X and Y relationship. Furthermore, the basis is philosophical, not scientific. Thus, the multiverse belief is remarkably similar to religious belief which also has no scientific underpinning.

We should readily admit that the prospect of a theory being false is a basic part of science. Only a falsifiable hypothesis can be considered scientific. The multiverse theory contains non-stop postulates that can be neither proven nor disproven, and rightly belongs in the metaphysical realm. That's not to say that parallel universes could not exist, only that there is zero evidence to support that it's scientific.

COSMOLOGY OF ANCIENT INDIA

It's appropriate at this point to bring up the research on ancient Hindu cosmology I cited in Chapter Three. Dr. Carl Sagan stated that a day of Brahma equals 8.64 billion years. The age of the universe by modern scientific reckoning is equivalent to only two days of the Hindu universe!

Their time scales are greater than ours? According to the ancient Sanskrit texts, the universe exists for 311 trillion years. Then it comes to an end, they claim, because nothing physical is permanent. The astronomers of antiquity not only calculated the birth of the cosmos, but also its demise. Ironically, these figures exceed the reckoning of modern science. But that's not the end of the story, because another cycle of 311 trillion years begins again. It's the endless cycle of the Vedic universe.

Paul Steinhardt, co-author with Neil Turok, of *Endless Universe: Beyond the Big Bang,* postulates that eventually our universe will reverse its expansion and begin contracting—the so-called Big Crunch. The Big Bang "repeats periodically every trillion years or so," Steinhardt asserts, "instead of being a one-time event."

This modern proposal is factually the original idea of ancient Indian astronomers as recorded in those Sanskrit texts that Carl Sagan informed us about. So Steinhardt is not a follower of the Big Bang, but rather the Big Bounce. Steinhardt is now in the cycle business—the cycles of the universe.

You may have noticed that the ancient astronomers gave an exact figure, 311 trillion years, while the modern version is vague, "every trillion years or so".

My research also revealed detailed descriptions of a multiverse in the ancient Sanskrit texts, described as innumerable bubble universes floating in a causal ocean.

Each bubble universe has its own span of existence independent of other super bubbles. All these cosmic bubbles are governed by innate physical laws unique to each, and thus not the physics that we know. These cosmic bubbles are so numerous that each one is compared to a mustard seed in a huge bag full of mustard seeds.

In spite of this written evidence, prominent authors like Michio Kaku and Bryan Greene promote parallel universes in a multiverse as the next big thing. Because they cite no reference, as Carl Sagan did, they must be unaware that astronomers of ancient India described the workings of a multiverse thousands of years ago. The Sanskrit texts are now translated into English, so if I can access the translations why can't they?

Another problem is that cosmologists refer to theories about parallel universes when in truth they are merely ideas. To be elevated to the status of a theory every hypothesis must have data as evidence to support and substantiate its ideas. Without data, the claim to a theory is simply a bluff.

I invite Dr. Tegmark, Dr. Kaku, and Dr. Greene to contact me regarding the parallel universe explanations from the cosmology of ancient India.

Colliding Particles

To validate that any kind of parallel universe exists, there must be some evidence. Be they large, tiny, or invisible, physicists must locate physical indications of extra dimensions that supposedly connect us to other worlds.

To accomplish this, scientists using the Tevatron particle collider at Fermilab are conducting extraordinary experiments with the expectation of proving that even the wildest kind of parallel universe is actually out there.

Joe Lykken is optimistic of success. "Right now, this detector is looking for evidence of extra dimensions. If there are extra dimensions of a certain size and shape, this experiment will find them."

The best chance to discover extra dimensions seems to come from smashing microscopic particles together at super speeds. But what are they looking for? The Fermilab physicists are looking for the particle that they believe carries gravity—the graviton.

Although the graviton is only a theoretical mathematical entity, Joe Lykken speaks of it as if it was real. "It's the particle that carries gravity, which we think knows about all the extra dimensions of space. So we think the gravitons, if we can produce them at high energies, should actually move off into the extra dimensions."

The Tevatron Collider shoots high energy protons one way and high energy anti-protons going the other way around a four mile long super-enforced steel-encased ring almost at the speed of light.

The collision ring lies deep underground for safety reasons and to protect the sensitive experiments from surface interference. At two places in the ring the particles smash together. The mighty collision annihilates the particles and produces an intense ball of pure energy.

Scientists hope that new particles will appear which might disappear into the extra dimensions. Until that happens, all bets are off. It's only theoretical at this point in time.

"It's tricky and complicated because you're looking for nothing," Lykken admits. "You're looking for something that has disappeared. It's very very rare that you make a really exotic particle, so we literally have to collide them billions of times in order to find the very rare event where you make something like a particle that disappears into extra dimensions."

The obvious question is how will scientists know that a graviton has moved off into an extra dimension if there is nothing to see?

"The way that you tell that," Lykken explains "is by reconstructing everything else that happened in this messy collision and then say, *Oops, there's some energy and momentum that's missing here.* So we call this a missing energy search."

The missing energy search leads us to an inescapable conclusion. Gravity is the smoking gun in the hunt for a theory to verify the existence of parallel worlds. Once again we find that solving the mystery of gravitational force will clear things up substantially.

Determined physicists believe they are now on the brink of uncovering the ultimate mysteries of the universe. But as history reveals, they have always had this belief.

SCIENCE FACT & MEDIA FICTION

Can we be certain there are parallel universes? Aleksey Fillipenko is optimistic. "At least some kinds, almost certainly, exist. Maybe the other kinds also exist. We just don't know."

Of course, "almost certainly" simply means we have high hopes.

Since all types of parallel universes are beyond any possibility to be observed theorists are faced with a real problem: how to actually validate the theory? They ignore the fact that this kind of science lacks testability. It has nothing to do with experiments that can be done in a laboratory or with data that can be detected by telescopes or microscopes. It even lacks the ability to be proven false. It has more in common with science fiction or, to be frank, religion.

But the scientists who support the multiverse concept are not acid heads. They are respectable professors driving fine cars, living in fine homes, and dining at fine restaurants. Perhaps they feast at diverse banquets in multiple parallel universes offering infinite menus.

Despite its proponents, M-theory has many outspoken critics both from within and from outside the scientific community. This is due to its infinite roulette wheels multiverse script.

Commenting on string theory in general, Richard Feynman remarked: "I am an old man now, and these are new ideas, and they look crazy to me, and they look like they're on the wrong track...I do feel, very strongly, that this is nonsense."[73]

When a physicist of Feynman's stature states that these ideas are nonsense, we should take it seriously. Reputable philosophers point out the fallacious thinking of multiverse advocates who, they say, abuse probability theory by theorizing beyond logic and reason.

Clearly, the multiverse scheme has two strikes against it. It's an unnatural explanation because it is based entirely on chance. There is no real evidence to support its existence, and there are no experiments that can test for it.

Even conservative string theorists deny the validity of infinite landscapes in infinite worlds. They argue that future developments will expose these assumptions as a mirage.

Nobel Prize winner David Gross is the leading voice of discontent among string theorists opposed to the multiverse idea. He anticipates that one elegant grand unified theory will finally emerge that corresponds to our observable universe, without the need for flights of fancy.

Physicist Paul Steinhardt of Princeton University considers the multiverse concept distasteful. "This is a dangerous idea that I am simply unwilling to contemplate."[74]

Does the multiverse allow for Santa Claus to exist in some parallel universe where he delivers gifts to every house in a single night with flying reindeer?

The fact remains: parallel universe ideas hover between

---

[73] P. C. W. Davies and J. Brown, *Superstrings - A Theory of Everything*, Cambridge University Press, 1998, pp. 193-194
[74] Paul Davies, *The Goldilocks Enigma*, Penquin Books, London, 2007, p. 194

Parallel Universes

science and fantasy. Within the cosmology community many persons do question whether the multiverse concept can be called science because, even in principle, it can't be verified or falsified by experiment, observation, or data.

Other physicists claim that the multiverse is rooted in respectable science and can be tested indirectly. Can *indirect* evidence verify the ideas?

According to expert mathematicians, maths can be worked to establish practically whatever they desire. Therefore, M-theory claims unlimited worlds with infinite possibilities from the sublime to the absurd. Even fake universes and simulated universes are included which might mean that even our observable universe may be fake.

Dr. Paul Davies assures us that predictions of fake universes outnumber predictions of real ones, "leading to the bizarre conclusion that the observed universe is probably a fake, and so its physics cannot be taken seriously anyway."[75] If this is the case, can M-theory be taken seriously?

The inescapable conclusion is that indirect unobserved evidence can never be considered reliable, especially when it relates to infinite possibilities. Some may be evidence of an illusion, or more to the point, they may create necessary illusions.

Nor can M-theory deny teleological and design conclusions. They are also likely candidates in a scenario with infinite possibilities. Unwittingly, M-theory has qualified the traditional creation story to assume scientific respectability.

In a recent *New Scientist* article Michio Kaku candidly voiced his concerns about where all this was headed. "If string theory itself is wrong, then millions of hours, thousands of papers, hundreds of conferences, and scores of books (mine included) will have been in vain. What we hoped was a 'theory of everything' would turn out to be a theory of nothing."[76]

---
[75] Paul Davies, *The Goldilocks Enigma*, Penquin Books, London, 2007, p. 299
[76] Michio Kaku, *New Scientist*, April 16, 2005

Joe Lykken notes: "There have been periods of many years where all of the smart people, all of the cool people, were working on one kind of theory, moving in one kind of direction, and even though they thought it was wonderful it turned out to be a dead end. This could happen to string theory."

Could string theory be wrong? Michael Duff admits it's possible. "Oh yes, it's certainly a logical possibility that we've all been wasting our time in the last 20 years and the theory is completely wrong."

A well-known adage can be updated for today's cosmology: One observation is worth a thousand theoretical equations. The valiant efforts of Tevatron notwithstanding, an observation is sorely missing.

There is no conceivable way to detect anything as tiny as strings, and thus no way to experimentally validate or invalidate string theory. Moreover, the theory needs extra dimensions of space to work; dimensions that can never be detected.

I wonder whether infinite possibilities actually exist, or whether the idea creates its own realism. The mathematics is the only thing scientists can hold on to.

The bottom line: if we can't test a theory it's not science. Therefore, faith in string theory is like faith in religion where the rituals are the only thing you can hold on to.

Paul Davies sums it up succinctly: "Given that the goal of string theory is to unify nature, it would seem to be a step backwards if the theory predicts a vast number of alternative worlds."[77]

The terms that define the new physics—like compactification of undetected dimensions, renormalizability, and bubble nucleation—do not define real world events. These are theoretical terms that physicists manufactured in an attempt to explain the unexplainable.

---

[77] Paul Davies, *The Goldilocks Enigma*, Penguin Books, London, 2007, p 128

To be fair, however, the renormalization idea also belongs to the Standard Model. It's how Feynman was able to do away with many infinities in his quantum electrodynamics theory, which gave very precise predictions. For that reason, it's considered one of the more successful theories in the history of physics.

In contrast, string theory remains a philosophy that is incompletely formulated and impossible to verify through experiments. If a theory can never be verified or even falsified, why would it become popular? Joe Lykken offers a possible explanation.

"What string theory does is it holds out the promise that, look, we can really understand questions that you might not really have thought were scientific questions; questions about how the universe began; why the universe is the way it is at the most fundamental level. The idea that a scientific theory that we already have in our hands could answer the most basic questions is extremely seductive."

Sir Roger Penrose, however, disagrees. He considers M-theory as merely "a collection of ideas, hopes, aspirations; it's not even a theory."

Leonard Susskind of Stanford University clarifies that physicists don't question if their ideas are a part of metaphysics, or philosophy, or religion. They just say, "Let's follow the logic." But we have shown that the reasoning is not logical. Even acclaimed physicist Richard Feynman feels, very strongly, that it's nonsense.

As far as gravity is concerned, there is still no explanation beyond theoretical concepts of undetected parallel universes, eleven dimensions, and the ever-elusive graviton. Yet we have Stephen Hawking's claim on national TV: "Gravity and quantum theory cause universes to be created spontaneously out of nothing." Very interesting, considering that gravity itself is still not clearly understood.

We already learned in Chapter Three that stable elliptical

orbits are only possible in three dimensional space. What kinds of orbits are possible in eleven dimensional space?

It's quite clear that the cause of gravity and the explanation of gravitational force is still a mystery for the modern creation story. And it's not a stretch to conclude that modern cosmologists and physicists have bequeathed to us the most fantastic theories, based on the wildest speculations, ever dreamed up by grown men. And they propose them on national TV!

As the day draws to a close our attorney summarizes the facts that have been clarified in today's session. In 300 years, from Newton's laws, Einstein's Relativity Theory, Quantum Mechanics, and String Theory, there is still no complete scientific understanding of how gravity works, what it's comprised of, and where it's located.

Suddenly, our lawyer drops a bombshell on the courtroom.

"Tomorrow," he claims, "we're going to explore the Santa Claus story. You may wonder, *why in the multiverse would I do that?* Well, the answer is quite simple—because it may prove to be relevant for understanding modern cosmology. Find out in tomorrow's session."

11
The Santa Claus Culture

Jolly old Santa embodies some of our higher ideals: childhood purity and innocence, selfless giving, and steadfast love.

He also reflects modern western culture. He is exuberant but over-weight; ubiquitous but highly commercial. Indeed, Santa has been tainted by some of society's greatest ills: corporate greed, materialism, and domination by the media. Nowadays, Santa has a lot more baggage than just his toys!

Why does society give children a fictitious story and present it as real? Most people say it's a harmless charade to get kids to behave. In return for good behavior kids are rewarded with nice presents.

Unfortunately, it involves falsifying facts and playing with a child's belief system. Is it a good idea for society to deliberately misguide children? Couldn't good boys and girls receive gifts from their parents or role models?

In court today our attorney plans to unravel the issue. "I want to look at an entertaining digression for a few minutes," he explains, "then we'll get back to the serious business of a trial."

The Santa Claus Myth

An editorial in the September 21, 1897 issue of the *New York Sun* asked: "Is There a Santa Claus?" The famous reply, "Yes, Virginia, there is a Santa Claus," comes from that editorial. It's now an enduring part of the Santa Claus culture in both Canada and the United States.

The movie, *Miracle on 34th Street*, gives credence to children that Santa is real. The film tells the story of an elderly gentleman, Kris Kringle, who claims to be Santa Claus. Most people are kindly towards Kris, but in almost every story there's a villain.

Before long, Kris is defending himself in a court of law for impersonating Santa Claus. It's late December, and the US Postal Service has a massive amount of mail sitting undelivered all addressed to Santa Claus. Someone at the Post Office has a bright idea.

Suddenly, in the middle of the trial, postal workers deliver to the courtroom sack after sack bulging with letters addressed to Santa Claus. Seeing that a government agency has accepted Kris as Santa Claus, the judge rules in his favor. It's a happy ending which gives Christmas cheer to all. For kids, it's a confirmation that Santa is real.

During the holiday season hundreds of thousands of letters are addressed to Santa Claus at the North Pole. What does the United States Postal Service do with all these letters? They direct them to the town of North Pole, Alaska, where a tourist attraction called 'Santa Claus House' is established.

The US Postal Service uses the town's zip code, 99705, as the advertised postal code for Santa. Every letter with a return address is answered. On the envelope is a North Pole postmark.

"Yes, Virginia, there is a town called North Pole." Only in America? Not by a long shot.

In Canada, Santa also lives at the North Pole but now it's

within Canadian jurisdiction. His address is: Santa Claus, North Pole, Canada, HOH OHO. (In other words, HO HO HO, Santa's jolly laugh).

On December 23, 2008, Canada's minister of Citizenship, Immigration and Multiculturalism, Jason Kenney, formally awarded Canadian citizenship status to Santa Claus. The *Toronto Sun* headline read: 'Santa Claus declared a Canadian citizen.'

The story was also broadcast on TV. In an official statement Kenney stated, "The Government of Canada wishes Santa the very best in his Christmas Eve duties and wants to let him know that, as a Canadian citizen, he has the automatic right to re-enter Canada once his trip around the world is complete." When children see this, they're convinced that Santa is real.

In Europe, each Nordic country claims Santa's residence to be within their territory. Norway claims he resides in Drobak. In Denmark, Santa lives in Greenland. In Sweden, the town of Mora has a theme park named Tomteland, and Stockholm's national postal terminal in Tomteboda receives children's letters for Santa.

In Finland, Korvatunturi has long been known as Santa's home. Two theme parks, Santa Claus Village and Santa Park are located near Rovaniemi. Santa Claus Village welcomes 300,000 visitors a year, with 70,000 visitors in December alone.

In Beijing, China, expatriate and local children receive their Santa letters from Finland. All letters with a return address get answered along with a Santa Claus Village postmark. This is organized under the auspices of the China Trade Commission, the Finnish Embassy International Post Office, Santa Claus Village in Rovaniemi, and the People's Republic of China Postal System's International Post Office in Beijing.

In 2009, Canada Post received 1.10 million letters from children in 30 different languages, including Braille. In addition, 39,500 e-mails and on-line request forms were answered by physical mail.

Since 1982, over 13,000 Canadian postal workers have volunteered to write replies to kids.

Santa Claus in Finland receives over 600,000 letters every year from over 198 different countries. Although he lives in Korvatunturi, the Santa Claus Main Post Office is situated in Rovaniemi near the Arctic Circle. His address is: Santa Claus Main Post Office, Santa's Workshop Village, FIN-96930 Arctic Circle.

In 2009, the Brazilian National Post Service, Correios, answered almost two million letters from children.

The following countries in 2006 received and answered letters to Santa Claus: Germany (500,000), Australia (117,000), Canada (1,060,000), Spain (232,000), Finland (750,000), France (1,220,000), Great Britain (750,000), Ireland (100,000), New Zealand (110,000), Portugal (255,000), Slovakia (85,000), Sweden (150,000), Switzerland (17,863), Ukraine (5,019).

In Mexico and other Latin American countries, besides using the mail, children sometimes tie their letters to a small helium balloon and release them into the air so Santa can magically receive them.

My wife is familiar with this practice. She would also leave a letter for Santa in a shoe before bedtime. Years later, she found a letter written to Santa that her mother had saved:

Dear Santa Claus,

I want you to bring me five hundred thousand pesos. Thank you very much. I've been a very good girl.

(signed) Christiane

Her younger sister, Linda, had not yet learned to write, but she took a pencil and scribbled all over a sheet of paper. When mother asked, "What are you doing?" Linda replied, "I'm writing a letter to Santa Claus."

The culture and tradition built around the Santa Claus legend can be traced back to the 1820s. Santa's home at the North Pole traditionally describes a village inhabited by his helpers, which surrounds his home and workshop. A large group of magical elves and other supernatural beings live in the village along with eight flying reindeer. The elves make the gifts he delivers to good children at Christmas.

A poem titled "A Visit from St. Nicholas" was published on December 23, 1823, in *The Sentinel*, of Troy, New York. Today it is known as "The Night Before Christmas."

The poem describes Santa's many attributes like riding in a flying sleigh which lands on the roof of each house. Santa enters through the chimney carrying a bag full of toys.

One of the first artists to define the modern Santa Claus image was an American cartoonist, Thomas Nast. A picture of Santa illustrated by Nast appeared in *Harpers Weekly* in 1863. A color collection of Nast's pictures was published in 1869 along with a poem by George P. Webster which was titled *Santa Claus and His Works*.

In the mid-1800s, Santa Claus got married. In 1889, the poet Katherine Lee Bates popularized Mrs. Claus in the poem *Goody Santa Claus on a Sleigh Ride*.

The 1956 popular song "Mrs. Santa Claus" and the 1963 children's book *How Mrs. Santa Claus Saved Christmas* by Phyllis McGinley helped establish the character and role of Mrs. Claus in the tradition.

Coca Cola did an ad campaign in 1931 that featured Santa. This accelerated his worldwide fame and made him into a commercial star. His present status as a giant celebrity of the modern commercial Christmas tradition is now global.

The 1934 song "Santa Claus is Coming to Town" stated that Santa makes a list of good boys and girls throughout the world and delivers presents to them on the single night of Christmas Eve. He accomplishes this feat with the aid of the elves, who make the toys, and the flying reindeer: Dasher, Dancer, Prancer, Vixen, Comet, Cupid, Donner and Blitzen, who pull his sleigh.

Another addition to the legend is the 9th and lead reindeer, Rudolph the Red-Nosed Reindeer, who was immortalized in a song by Gene Autry.

How does Santa arrive before Christmas? In Canada, there is an annual Santa Claus parade in major cities. Kids and parents crowd the streets to see Santa ride in his sleigh with reindeer, usually on the last float. Now that he's in town, children can visit him at any Eaton's department store throughout the country, sit on his knee and ask for Christmas gifts.

In the US, Santa arrives at Macy's department store in New York City via Macy's Thanksgiving Day Parade. He waves from his sleigh on the last float, and later his court takes over a large portion of one floor in the store.

David Sedari, who formerly worked as an elf in Macy's Christmas display, is known for his satirical *Santa Land Diaries* that were broadcast on the radio and later published.

If Santa is ever detected to be fake, he will explain that he's not the real Santa but he's helping him out at this time of year. Most young children seem to understand that the real Santa is extremely busy around Christmas.

If all this isn't enough to convince a child that Santa is real, consider what takes place on Christmas Eve. A number of websites have been created over the years that actually track Santa Claus on his trip around the world. Among the most well-known are: NORAD Tracks Santa, the Air Services Australia Tracks Santa Project, the Santa Update Project, and the MSNBC and Bing Maps Platform Tracks Santa Project.

Santa tracking began in 1955 by accident. A mistyped number in a Sears Roebuck ad that told kids to call a "Santa hotline" turned out to be a number for the Continental Air Defense Command (CONAD). Colonel Harry Shoup was the Director of Operations at the time. He received the first call for Santa and responded by saying that Santa could be seen on the radar—his sleigh was now heading south from the North Pole.

When Canada and the United States jointly created the North American Air Defense Command (NORAD) in 1958, a tradition began under the name NORAD Tracks Santa. Today, children can track Santa via the Internet and NORAD's website.

In the US and Canada local TV stations used to track Santa Claus via the weather man. Then the Weather Channel built upon these local efforts by providing a national 'Santa tracking' effort called 'Santa Watch' in December 2000. It works in cooperation with NASA and the International Space Station.

Nowadays, most local TV stations rely upon established 'Santa tracking' efforts. When kids see this on TV, they are convinced. "This is for real!"

In the US and Canada, children traditionally leave Santa a glass of milk and a plate of cookies. In Britain and Australia, he can get a glass of sherry and mince pie. In Sweden, youngsters leave rice porridge. In Ireland it's popular to give Santa hot milk or Guinness, and Christmas pudding or mince pie.

Many websites are now devoted to Santa Claus and purport to keep tabs on activities in his workshop. The websites also have electronic addresses so children can e-mail Santa. Most websites have volunteers that become 'elves' to answer Santa's e-mail.

Besides offering holiday-themed entertainment, 'Santa tracking' websites are ultimately planned to inspire kids to think about science. Their goal is to encourage youngsters to take an interest in science and learn how space technology and exploration are an important part of daily life.

By the end of the 20th century, the reality of mass mechanized production became fully accepted by the public. That shift reflected the modern image of Santa's residence. It's now portrayed as a highly advanced production and distribution facility, equipped with the latest manufacturing technology, all overseen by the elves. Santa and Mrs. Claus are executives and managers. An excerpt from a 2004 trade magazine illustrates this depiction:

> "Santa's main distribution center is a sight to behold. At 4,000,000 square feet (370,000 square meters), it's one of the world's largest facilities. A real-time warehouse management system (WMS) is of course required to run such a complex. The facility makes extensive use of task interleaving, literally combining dozens of DC activities (put-away, replenishing, order picking, sleigh loading, cycle counting) in a dynamic queue...the DC elves have been on engineered standards and incentives for three years, leading to a 12 percent gain in productivity...
>
> "The WMS and transportation system are fully integrated, allowing the elves to make optimal decisions that balance transportation and order picking and other DC costs. Unbeknownst to many, Santa actually has to use many sleighs and fake Santa drivers to get the job done Christmas Eve, and the transportation management system (TMS) optimally builds thousands of consolidated sacks that maximize cube utilization and minimize total air miles."

After reading this, tourists from a foreign culture who have never heard of Santa Claus could reasonably conclude that he was a real person.

Today the charade is massive and worldwide. The hoax is so intricate and so expertly portrayed that it's almost impossible for youngsters to see through the ruse until they mature.

Criticism comes from various groups that the Santa deception is not a simple lie, but a complicated series of very large lies that presents a certain reality which is factually false.

Writer Austin Cline posed the question: "Is it not possible that kids would find at least as much pleasure in knowing that parents are responsible for Christmas, not a supernatural stranger?"

How does the Santa experience affect the belief system and sense of trust in children, especially when they mature into adults?

Jaqueline Woolley noted in a December 23, 2006, *New York Times* article, that children depend on adults to provide reliable information about the world. Yet adults dupe them into believing Santa Claus is real. This reality is confirmed by friends, books, TV, movies, Santa Tracking web sites, and a Santa in the store that the kids talk to. It's also validated by hard evidence: half-eaten cookies and empty milk glasses by the tree on Christmas morning.

Wooley suggests that children do a good job of evaluating Santa's reality, but adults do a better job of duping them. She proposes that "kinship with the adult world" causes children to not be angry that they were lied to.

In the 2008 film *Religulous*, however, there's a scene with Bill Maher, his mother and sister. The mother reveals that Bill was so angry at her when he found out there was no Santa Claus.

Ironically, many kids grow up and play the same hoax on their own children because society is so ingrained with it on TV, in the movies, Santa in the stores, songs on the radio, web sites, letters from Santa stamped by the post office, and on and on. So the fault may not lie entirely with the parents. It's a societal charade and parents are as enraptured as the kids.

Why do we invent a complete culture to deceive our kids? Psychologist Tamar Murachver sees no harm in the Santa Claus belief because it's a cultural lie, not a parental lie; therefore it doesn't undermine parental trust.

This is not always the case. A friend wrote me about his experience.

"I would speak out about how I felt at discovering Santa was a lie. It was some older boys at school, and they made me feel terrible, and foolish. After that my faith in my parents was never the same. Their credibility was trashed. Of course, I went along to get the gifts and have the 'fun' of Christmas, but later when I was to accept Jesus as my savior, it was like, 'Yeah, right. Let me take a rain check on that one.' I was disenfranchised to say the least. To this day the foolishness is going on."

When my wife was a child, she really believed Santa would come down the chimney, even though there was no chimney in her house. She was told that he comes down the chimney, so why would she doubt her parent's story? His sleigh and reindeer remained waiting for him above the house because they could fly. He brought the gifts and then quickly departed.

She and her younger sister would wake up early next morning and tip-toe down the stairs to see if Santa was still there. They would peek into the room and upon seeing the gifts under the tree were delirious with joy.

At the age of eight she was so excited about Santa Claus coming that one girl at school made fun of her. "You still believe in those things? It's your parents."

It was a shock that swept away her excitement. She was really hurt. At home she asked her mother if it was true. Hearing that it was indeed true, she began to cry. It was such a disappointment. She had imagined how Santa traveled all over the world doing a superhuman task because he loved children. And now it was all false.

"I really felt like a fool," she revealed. "I was fooled by people that I trusted; my parents. And my mother made it even more fantastic by saying things like, 'I just saw Santa Claus and he said he was going to bring you so many things because you were such a good girl.' But none of it was true."

The Santa Claus Culture

Once you've convinced your kids that Santa is real, it's very difficult to say, "He's not real." My wife's mother had a lot of explaining to do.

Parents have to explain the entire hoax: Santa in the store whose lap the kid sat on, how the presents appeared, and how the adults played along that it was Santa when they knew it wasn't. They have to explain that there is no North Pole address, even when the kids receive letters from Santa with a North Pole postmark. The mass media and local TV stations are in on the scam.

Once we weave such a detailed story, admitting that it's untrue is intolerable. The only hope a parent has of escaping from the deception is to wait until the kids hears it from someone else. Then they can take the role of the consoling parent and offer reasons why they played along with the charade.

Educators are quick to point out that a letter to Santa is often a child's first experience of correspondence. Written and sent with the help of a parent, children learn how to write letters. Advocates for the tradition say the Santa Claus culture is a useful model to show that other beliefs in the supernatural might be equally false. In other words, don't blindly accept any belief.

And what about blindly accepting the science creation story? If society can spin a web so all-encompassing that no child could ever figure out it was a fairy tale, is it possible to spin a web to deceive adults?

Don't some governments invent a total fabrication and present it as factual to influence their citizens a certain way?

And don't multi-national corporations spend millions of dollars on mass media advertising to convince people they must have a certain product?

What about the medical industry? Don't they try to persuade people that their prescription drugs are safer and better than cures from nature?

Drugs can alleviate symptoms quickly, but they never provide a lasting cure. A cure is bad for business. A cure means you lose a customer. And don't forget the side effects. Now that's good for business—repeat customers.

There is always an ulterior motive or a hidden agenda in the minds of businessmen and corporations. They don't want to lose repeat sales and lifelong customers.

How does all this relate to cosmology?

On a societal level, the Big Bang origin of the universe is as widespread and as accepted by adults as the Santa Claus story is accepted by kids.

Geoffrey Burbidge, Professor of Physics at the University of California, San Diego, concurred with this conclusion:

"Big Bang cosmology is probably as widely believed as has been any theory of the universe in the history of Western civilization. It rests, however, on many untested, and [in] many cases, untestable assumptions. Indeed, Big Bang cosmology has become a bandwagon of thought that reflects faith as much as objective truth."

When 96 percent of the matter and energy in the cosmos is unknown to scientists, cosmology can only teach about the 4 percent that is known. How can we be certain about what we understand of the universe on the basis of a paltry 4 percent? With a full 96 percent of the cosmos mysterious and unknown, why do cosmologists promote their creation story to the public by elevating it to the status of scientific knowledge?

Most science is on solid ground, but in cosmology it's more like thin ice. Indeed, some of the wild theories proposed nowadays do tend to sound like science fiction. Especially, since there's no testable data to back it all up. Of course standing on thin ice is a normal situation for any developing field of research.

My point, however, is that the creation story of physics is taught in educational institutions as scientific, whereas the traditional God created universe is considered a bronze-age superstition. Yet, they both seem to stand on the same platform—faith without evidence.

I'm surprised that the modern creation story has elements in common with the Santa Claus story. Particularly, as cosmologists promulgate invisible parallel universes with infinite possibilities and copies of you and me.

Our attorney turns to the jury and says, "We're now going to examine the Big Bang theory to see whether it still holds water with the accumulated data of the 21st century."

"Let that be the focus of your investigation after lunch," says the Judge as he declares a one hour recess.

12
Hubble's Dilemma

For thousands of years, people looked up at the sky with awe and wonder. In every culture the prevailing view was that the cosmos must have been created by some kind of supernatural being, variously described as a god.

So far we've been looking at evidence of scientific efforts throughout history. But what does 21st century science tell us specifically about the cosmos? This is the topic that will now be presented to the jury.

Expanding Universe

About a thousand billion galaxies are currently in range of modern telescopes. Each galaxy contains a hundred thousand million stars. What lies beyond the range of our telescopes, of course, is unknowable.

The current understanding in cosmology is that the space between the galaxies is expanding. As time moves forward the galaxies will be carried further and further apart from each other. This cosmic expansion proceeded from the Big Bang. We are told that it's not an explosion in space; rather it's an explosion of space itself. Everything we observe with our instruments is embedded within the explosion.

When we consider that the expansion of space exceeds the speed of light, clearly, we will never know or never observe anything beyond our own limited cosmological horizon.

This begs another question: Is there anything outside the explosion? The answer from science is that nothing exists outside the explosion because there was no space or time before the Big Bang. The universe simply expanded from an infinitesimal size to an infinite size. Going back to the remote past, presumably to the beginning, our location today corresponds to a location within that infinitesimal singularity.

The conclusion is that the Big Bang happened at all places at the same time. That's the working hypothesis of cosmology. But it might just be an arbitrary precept.

Skeptics may ask, how did the singularity come into existence before it generated the cosmos? Moreover, how did the entire infinite matter and energy get squeezed into a subatomic size? Finally, what caused the singularity to change its state to bring forth what we see today as the universe?

The new multiverse idea posits that all possibilities exist only within the expansion of space. This means we can never observe or test what lies outside because we're enclosed within the expansion. Although we can't even see the farthest reaches of the universe, science concludes that nothing exists outside. This may be a blunder, quite simply, because it's not provable, so it becomes like dogma.

Most scientists today believe the universe is infinite. According to the Big Bang cosmological model, the universe had infinite density, infinite mass and infinite temperature, when we trace it back to the initial moment the subatomic singularity exploded. As far as cosmology is concerned, the singularity contained all space and time in an infinitesimal volume.

This rationale is the entire basis for the Big Bang theory. Does the idea represent reality? Nobody knows for sure. Basically, it's a belief supported by mathematics, which can be manipulated, as we learned in Chapter Eight.

When an idea can't be tested, any one answer has the same probability of being right as any other answer arrived at by another method. Such manipulation of figures, without the backing of observation, is not really science, as various physicists themselves have lucidly pointed out.

Some people will argue that we can test the Big Bang model via different predictions of the theory using various indirect methods of observation, like the CMB, the Hubble Constant, etc. This does rule out most crazy theories, and it has done so in the past. So let's examine how these predictions support the current understanding of cosmology.

Static Universe

At the beginning of the 20th century we had a totally different idea of the cosmos. Science taught that our Milky Way galaxy was the entire universe. Every physicist believed in a static universe that always existed.

Since the universe was considered to be eternal, there was no need to question where all the matter and energy came from. It was always there. This was the scientific world view.

Then, Einstein published his two brilliant theories: Special Relativity in 1905 and General Relativity in 1915. It soon became clear that his mathematics indicated the cosmos could be either expanding or contracting. This disturbed him because he was convinced the universe was static. Nobody believed the universe was expanding a hundred years ago.

So what did Einstein do? He dismissed the idea of an expanding universe and adjusted his equations to overcome what he perceived as a problem. He fudged the formula by adding a cosmological constant to keep the universe static. By mathematical manipulation he achieved the result he wanted.

Georges LeMaître, a Roman Catholic priest from Belgium, was also a professor of physics. In 1927 he published a paper which demonstrated that a linear velocity-distance relationship between galaxies supported a model of an expanding universe based on Einstein's general relativity equations.

Running the 'movie' of an expanding universe backwards into the remote past, LeMaître proposed that all matter and energy of the cosmos would be compacted. Going back far enough, to the beginning of time, we would reach an event when it all began. He called this cataclysmic moment of creation the primeval atom.

In LeMaître's primeval atom astronomers were faced with a beginning; a consensus between the traditional creation story and the rigors of science. LeMaître not only proposed that the universe was actually expanding he also posited a value for the rate of expansion. The implication was a non-static universe.

The idea that the universe was expanding and had a beginning was a radical concept that nobody was willing to consider. Every scientist at the time took it for granted that the universe was static. Very few took the expanding cosmos seriously.

Hubble's Observations

Data to support the expanding universe idea came from Edwin Hubble.

In the early 1920s, Hubble was studying the skies with the brand new 100 inch telescope at the Mount Wilson Observatory in California. He was able to determine the distances and velocities of neighboring galaxies, demonstrating that they were separate and distinct from our own. That's how he discovered that the Milky Way galaxy was not the entire universe, but rather just one among many galaxies.

Back then, most scientists in the astronomical establishment opposed this idea. Despite the resistance Hubble presented his findings at the January 1, 1925 meeting of the American Astronomical Society.

In 1929, Hubble published a paper proposing that the universe was indeed expanding. Careful examination of his observations revealed that every galaxy was receding from Earth in every direction. He calculated a more accurate expansion rate than LeMaître—the Hubble Constant.

The observational data from his telescope was an earth shattering revelation for his colleagues. His paper challenged the scientific understanding of the universe. The prevailing view was suddenly turned upside down.

In 1931 both Einstein and LeMaître paid a visit to Hubble. After Einstein saw the telescopic data he admitted that his cosmological constant was the "biggest blunder" of his career.

LeMaître and Einstein Historical Photo universetoday.com

Einstein's example illustrates how a preconceived idea, or bias, colors our work and our understanding. It is essential to remove bias from research in order to discover true causal relationships. Bias refers to known or unknown influences that affect the result. Researchers must do their best to account for and control such influences.

Science requires that even unknown biases must be accounted for and controlled. Otherwise, theories can never conclusively establish causal relationships beyond some hidden variable influencing the result.

To keep the universe static, Einstein added a fudge factor to his equations to get a result that was acceptable to the scientific establishment. Although he didn't like the constant, Einstein added it to conform to the accepted view. But that wasn't his biggest botheration. If he had been alert he would have received the credit for proving an expanding universe and even pocketed another Nobel Prize.

Due to the work of Hubble and LeMaître, Einstein finally accepted the theory of the primeval atom. Today we call it the Big Bang. These three men were the originators of 20th century cosmology.

But there was still a cadre of scientists who did not accept the new theory.

Cambridge cosmologist Sir Fred Hoyle took issue with this model of the universe. He argued that an expanding universe doesn't have to start with a bang. It could instead be expanding endlessly while generating additional matter to fill the empty gaps.

Hoyle proposed a new theory in 1948, the Steady State model. Accepting that the universe was expanding, he suggested that as galaxies moved away from each other new matter continuously appeared between the galaxies.

He decided to go on the air to inform the public that the question of how the universe began was far from settled. [As I am doing with this book]

On a radio program in 1949, Hoyle rejected the scientific validity of the primeval atom theory. He explained to his radio audience that he doubted the universe began with a bang:

> "Perhaps like me you grow up with the notion
> that the whole of the matter of the universe

was created in one big bang at a particular time in the remote past. What I'm now going to tell you is that this is wrong. Now this big bang idea seemed to me to be unsatisfactory even before detailed examination, for it's an irrational process that can't be described in scientific terms."

Thus, the "Big Bang" name was coined by Hoyle. Although he rejected the theory, the name caught on and stuck. Many physicists at the time acknowledged their preference for Hoyle's Steady State model.

Dennis Sciama: "I personally liked the theory because I thought it had a grand architectural sweep. It just seemed so grand to have a universe that didn't change in its large scale structure ever, and had no awkward initial moment. If you go backwards in time there's no increase of density and therefore no big bang. So you have this rather grand picture of a universe which is expanding but which stays the same in its overall properties for all time."

To bolster the Steady State theory, Hoyle proved in 1952 that all heavier elements are generated and spread throughout the universe by supernova explosions. Matter was continuously being produced as Hoyle had posited. The Steady State model gained popularity.

Of course the theory had no explanation how matter, and the energy needed to produce it, was first created overall. The belief that the universe always existed avoids the issue of how it initially got going, just as the belief in an eternal god avoids the issue of how he was created.

The advantage the Big Bang model had was addressing the problem of origin. It proposed an explanation how everything in the cosmos was created within the cataclysm of the Big Bang itself.

But for a long time physicists didn't like this theory. It was considered too controversial.

Cosmic Microwave Background

If there was a Big Bang might there still be heat radiation left over from that awesome explosion? Could it be detected within the cosmos?

These were the questions formulated by Robert Dicke of Princeton University in 1964. But before he could get an experiment going, Robert Wilson discovered what seemed to be "noise" coming from space. He called Dicke for advice in the spring of 1965.

Subsequently, it was confirmed that Wilson had discovered radiation coming from space which might be left over from the Big Bang. The discovery of this cosmic radiation of the universe, now known as Cosmic Microwave Background (CMB) radiation, caused a paradigm shift in cosmology research.

This discovery was a blow to the burgeoning Steady State theory. It was difficult to explain the source of the background radiation in a Steady State universe. And so the Big Bang theory became the accepted model of the creation of the universe. The Nobel Prize was awarded to Wilson and his colleagues in 1978.

The Penrose Hawking singularity theorem, which we discussed in Chapter Four, also helped to establish the Big Bang model. But there were still some questions that needed to be answered. How did the galaxies form?

Physics taught that matter coalesced in cooler and denser pockets to form the galaxies. Therefore, the early universe should contain slight irregularities in the signal. The radiation Wilson had detected didn't show any signs of that. If the Big Bang model was right, where were the irregularities?

George Smoot decided to look for the temperature fluctuations that would have allowed galaxies to form. Not an easy task. To prevent Earth's atmosphere from interfering and distorting the results, all the measurements would need to be taken from space. But NASA was reluctant to get involved.

Finally in 1989 NASA agreed to send a satellite into space for cosmology research. At last, Smoot was given his chance. NASA called the satellite the Cosmic Background Explorer or the COBE Project.

George Smoot: "It took a whole year's worth of data, 300 million observations, to be summed together before we got a map that started to show an interesting structure with irregularities."

The data map from COBE put the Big Bang theory on a solid footing. The 2006 Nobel Prize for Physics was awarded to George F. Smoot, UC Berkeley, and John C. Mather, NASA, for the discovery of the blackbody form and anisotropy of the CMB radiation.

The WMAP satellite was the successor to COBE. Launched in 2001, it yielded even better readings of the CMB temperature fluctuations. The full sky map of the universe was accepted as scientific proof that confirmed the birth of the universe and the formation of galaxies. At last, we were witness to the dawn of time. The Big Bang model was now irrefutable.

Or was it? The account you have just read is what they teach in astronomy and cosmology classes.

Our attorney now proposes to the jury that we revisit the history to see if there is any missing information.

Hubble's Dilemma

How did Edwin Hubble determine that galaxies were receding away from the Earth in all directions? He did it by studying the light from deep space.

Every light source has its own spectrum, the rainbow of colors that's visible when light is separated by a prism. The spectrum reveals two types of spectral lines: 1) emission lines that are light on dark, and 2) absorption lines that are dark on light. When the frequencies don't align correctly the spectrum is said to be shifted.

A redshift means the absorption lines are shifted towards the red end of the spectrum. The change in frequency is interpreted as a velocity shift caused by motion. A shift in red indicates the source is moving away from the observer. A shift in blue indicates the source is moving towards the observer.

Where does this idea come from? Just as the sound of a moving locomotive is shifted as it passes an observer, the Doppler Effect, similarly redshift must be caused by lightwaves being stretched in transit from the perspective of an observer. This "Doppler Shift" concept became the primary interpretation of galactic redshift for most of the 20th century.

By observing a redshift in the light of all objects from interstellar space Hubble naturally interpreted it as a Doppler effect related to Earth. It was the scientific explanation of his day, and therefore Hubble derived the recession velocity of objects by comparing their redshifts.

In 1929, he proposed his Redshift Distance Law: the greater the distance between any two galaxies, the greater their relative speeds of separation. Their velocity is proportional to their distance from Earth and other stellar bodies.

Physicists who liked the expanding universe idea had initially accepted the hypothesis based solely on Einstein's mathematics. The missing ingredient had been observational evidence. Hubble's work via the Doppler shift concept now provided an explanation for the expanding space paradigm to support the Big Bang model of creation.

Although the redshift factor came to be accepted as a measure of recession speed, Hubble was well aware that redshift interpretation was an arbitrary assumption based on the Doppler Effect of sound waves. There was no hard data, no factual evidence, to connect redshift of light waves with the Doppler shift of sound waves.

However, because it fit the prevailing worldview, Hubble accepted it. But his colleague, Milton Humason, did not. Although they had worked together for years, Humason was reluctant to accept redshift as a Doppler phenomenon without hard data. In his opinion, redshift was not observational evidence for an expanding cosmos.

In 1931, he published a paper expressing his views on the subject:

> "It is not at all certain that the large redshifts observed in the spectra are to be interpreted as a Doppler effect but, for convenience, they are interpreted in terms of velocity and referred to as apparent velocities."[78]

Humason was only willing to commit to the interpretation of "apparent velocities" for the sake of convenience. Cognizant of his colleague's doubts, Hubble also wanted to be certain that the measured redshifts of various bodies observed from Earth were true Doppler effects.

In the 1930s Hubble relocated to the much more powerful 200 inch telescope at the Mount Palomar Observatory. Having the world's most advanced telescope as his instrument, his examination of numerous light sources from distant stars was unsurpassed. His goal was to prove that redshift was indeed a Doppler effect.

It soon became clear to Hubble that if redshift was a Doppler effect, then every stellar light source was receding away from Earth in every direction. That would mean Earth must occupy a central place in the universe. But how could he en-

---

[78] "Velocity-Distance Relation Among Extra-Galactic Nebulae", *Astrophysical Journal*, 74, 1931

dorse such a conclusion?

The premise of a central Earth had been overturned since the 16th century. Interpreting redshift observations from his telescope as a velocity indicator was thus very upsetting to the current view. Consequently, he began to doubt his conclusion of 1929 that redshift indicated a Doppler effect.

Both Hubble and Humason were troubled by the redshift anomaly. Could redshift have some other interpretation besides a Doppler effect?

In 1937, Hubble put this point to the Royal Astronomical Society. If science accepts redshift to be a Doppler effect it causes a major problem:

> "The observations as they stand lead to the anomaly of a closed universe, curiously small and dense, and it may be added, suspiciously young. On the other hand if redshifts are not Doppler effects, these anomalies disappear and the region observed appears as a small, homogeneous, but insignificant portion of a universe extended indefinitely both in space and time."[79]

If Hubble admitted to redshift being a Doppler effect, then he would be forced to accept an Earth-centered universe which he describes as 'closed,' 'small,' 'dense' and 'young.' In this way, one of the world's great astronomers admitted that telescopic observations were not in line with views held by modern astronomy.

As early as 1934, Hubble confirmed his position at the Halley Lecture:

> "The field is new, but it offers rather definite prospects not only of testing the form of the velocity-distance relation beyond the reach of the spectrograph, but even of critically testing the

---

[79] E. Hubble, Monthly Notices of the Royal Astronomical Society, 97, 513, 1937

> very interpretation of redshifts as due to motion. With this possibility in view, the cautious observer refrains from committing himself to the present interpretation and prefers the colorless term 'apparent velocity'."[80]

Hubble had adopted Humason's term "apparent velocity" because he shared the same conclusion. Since there was no other explanation for redshift at the time, he continued writing about his concerns with the Doppler interpretation, without providing an alternative.

By 1936, Hubble revealed that his Mount Palomar telescopic data established the disturbing conclusion that if redshifts measure expansion rates, the resultant expanding model doesn't tally with observations.

> "...we find ourselves in the presence of one of the principles of nature that is still unknown to us today ... whereas, if redshifts are velocity shifts which measure the rate of expansion, the expanding models are definitely inconsistent with the observations that have been made ... expanding models are a forced interpretation of the observational results."[81]

Here was Hubble's dilemma. He had originally acknowledged the redshift explanation to substantiate the expanding universe idea of Georges LeMaître, and Einstein's theory of General Relativity. But now, the evidence suggested that redshift may not be a Doppler effect. If true, the data revealed a universe not governed by Einstein's equations or the modern Big Bang theory.

---

[80] "Redshifts in the Spectra of Nebulae", *The Halley Lecture*, May 8, 1934, Oxford: Clarendon Press, 1934, p.14
[81] E. Hubble, *Astrophysical Journal* 84, 517, 1936, p. 553

> "...if redshifts are not primarily due to velocity shifts, the observable region loses much of its significance. The velocity distance relation is linear; the distribution of nebulae [galaxies] is uniform; there is no evidence of expansion, no trace of curvature, no restriction of the time scale."[82]

Space "curvature" and "restriction of the time scale" are two of Relativity's basic tenets. Without the relation between redshift and velocity as a Doppler shift, Einstein's theory of Relativity was in trouble.

Clearly, "no trace of curvature" suggested a flat universe, one in which space did not curve. Understandably, Hubble was concerned about his original 1929 interpretation. He concluded that there must be some undiscovered principle of physics at work. With no equivocation, he was adamant that a redshift/velocity relationship was inconsistent with observable data.

Data from the most advanced telescope on earth prompted him to conclude that the expanding universe of Einstein's mathematical equations was "a forced interpretation of the observational results."

In defiance of the mathematical approach to cosmology, Hubble published in 1937 a book called *The Observational Approach to Cosmology*. The book clarified that his method was observational and not theoretical. He explained a Doppler interpretation of the redshift condition in this manner:

> "Such a condition would imply that we occupy a unique position in the universe, analogous, in a sense, to the ancient conception of a central Earth...This hypothesis cannot be disproved,

---

[82] Astrophysical Journal 84, 517 (1936), p. 553; and *The Observational Approach to Cosmology*, p. 63

> but it is unwelcome and would only be accepted as a last resort in order to save the phenomena. Therefore we disregard this possibility...the unwelcome position of a favored location must be avoided at all costs...Such a favored position, of course, is intolerable; moreover, it represents a discrepancy with the theory, because the theory postulates homogeneity. Therefore, in order to restore homogeneity, and to escape the horror of a unique position, the departures from uniformity, which are introduced by the recession factors, must be compensated by the second term representing effects of spatial curvature. There seems to be no other escape."[83]

If redshift is interpreted as a Doppler effect, the evidence coming from the world's most powerful telescope suggested a central Earth. That was not only "unwelcome" but it was a "horror."

Hubble considered that a geocentric universe was "intolerable." But even more alarming was the fact that a geocentric hypothesis could not be disproved. How to deal with this apparent geocentric situation?

Hubble concluded that it "must be avoided at all costs." The thought that science may have been wrong for three centuries was "intolerable." It was a "horror" that any scientist would want to avoid.

When confronted with evidence that contradicts the prevailing paradigm even the most gifted scientists have difficulty overcoming personal convictions. "For a man who has been honored, dishonor is worse than death."[84]

In his book, Hubble admits that the more data that came in,

---

[83] E. Hubble, *The Observational Approach to Cosmology*, Oxford, Clarendon Press, 1937, pp. 50, 51, 58

[84] Vedic aphorism from the *Bhagavad Gita*

the more his doubts about the Doppler interpretation of redshift increased.

> "But later, after the 'velocity distance relation' had been formulated, and Humason's observations of faint nebulae began to accumulate, the earlier, complete certainty of the interpretation began to fade."[85]

He could no longer commit to redshift being a Doppler shift, an actual velocity effect. To avoid the horror of a geocentric universe, he considered the term "apparent velocity" acceptable, even though the interpretation seemed to contradict his telescopic data, because no other explanation for redshift was known then. That's why he introduced the possibility that an unrecognized, or unknown, principle in physics might explain the redshift.

> "This explanation interprets redshifts as Doppler effects, that is to say, as velocity-shifts, indicating actual motion of recession. It may be stated with some confidence that redshifts are velocity-shifts or else they represent some hitherto unrecognized principle in physics…
>
> "Meanwhile, redshifts may be expressed on a scale of velocities as a matter of convenience. They behave as velocity-shifts behave and they are very simply represented on the same familiar scale, regardless of the ultimate interpretation. The term "apparent velocity" may be used in carefully considered statements, and the adjective always implied where it is omitted in general usage."[86]

---

[85] *The Observational Approach to Cosmology*, Oxford, Clarendon Press, 1937, p. 29
[86] *The Observational Approach to Cosmology*, Oxford, Clarendon Press, 1937, p. 22

Hubble left himself open. He didn't commit to the prevailing view but went along with it for the time being. At the same time, he continued to air his doubts because it was important for him to be absolutely certain.

Hubble's conclusions of whether redshift was related to velocity caused quite a stir. Yet he continued to maintain, "if redshifts are not primarily due to velocity shift...there is no evidence of expansion, no trace of curvature, no restriction of the time scale."

As far as science is concerned, experiments have to be made to ascertain the facts. Therefore, Hubble conscientiously performed various experiments to determine the nature of redshift. But all the data seemed to substantiate that redshift was *not* a Doppler effect of velocity.

> "Since the intrinsic luminosities of nebulae are known, their apparent faintness furnishes two scales of distance, depending upon whether we assume the nebulae to be stationary or receding. If, then, we analyze our data, if we map the observable region, using first one scale and then the other, we may find that the wrong scale leads to contradictions or at least to grave difficulties.
>
> "Such attempts have been made and one scale does lead to trouble. It is the scale which includes the dimming factors of recession, which assumes that the universe is expanding."[87]

Hubble had to concede that the observational evidence for an expanding universe "does lead to trouble." He was left with no choice based on the overwhelming data revealed by his telescope.

In order to maintain the validity of the prevailing worldview,

---

[87] E. Hubble, "The Interpretation of the Redshifts", pp. 108-109

however, he tried to accept a different understanding. Therefore, he took a shot at Relativity and Inter-nebular Material [today's Dark Matter]. He expressed his opinion of Einstein's General Theory of Relativity succinctly:

> "Thus the theory might be valid provided the universe was packed with matter to the very threshold of perception. Nevertheless, the ever-expanding model of the first kind seems rather dubious. It cannot be ruled out by the observations, but it suggests a forced interpretation of the data.
>
> "The disturbing features are all introduced by the recession factors, by the assumption that red-shifts are velocity-shifts. The departure from a linear law of red-shifts, the departure from uniform distribution, the curvature necessary to restore homogeneity, the excess material demanded by the curvature, each of these is merely the recession factor in another form…if the recession factor is dropped, if red-shifts are not primarily velocity-shifts, the picture is simple and plausible. There is no evidence of expansion and no restriction of the time-scale, no trace of spatial curvature, and no limitation of spatial dimensions. Moreover, there is no problem of inter-nebular material."[88]

Hubble's complaint is related to today's issue about dark matter. He was now confident that redshift had nothing to do with a Doppler effect. So why do modern scientists cling to the existence of dark matter when they don't even know what it is? The answer is simple; without dark matter Einstein's field

---

[88] Edwin Hubble, *The Observational Approach to Cosmology*, Oxford, Clarendon Press, 1937, p. 63

equations will not work. In 2006, scientists from the Max Planck Institute in Germany came to this same conclusion.

> "Dark Matter is needed if one assumes Einstein's field equations to be valid. However, there is no single observational hint at particles which could make up this dark matter. As a consequence, there are attempts to describe the same effects by a modification of the gravitational field equations, e.g. of Yukawa form, or by a modification of the dynamics of particles, like the MOND ansatz, recently formulated in a relativistic frame. Due to the lack of direct detection of Dark Matter particles, all those attempts are on the same footing."[89]

If Einstein's field equations are doubtful without dark matter, does the same hold true for the Big Bang model that took birth from the equations? This issue has become more dramatic in astrophysical circles because dark matter has never been detected. Like the Doppler interpretation of redshift, it remains an untested hypothesis to support General Relativity.

In spite of all this, the redshift/velocity ratio is taught in physics courses as Hubble's Law. Originally, it was attributed to "Hubble-Humason." But when it became clear that Humason would not commit to the Doppler interpretation, his name was dropped. Milton Humason paid a price for his noncompliance.

Ironically, Hubble shared the view of his colleague yet his name was retained. If he also didn't commit to the prevailing redshift interpretation why do we have Hubble's Law? The reason is that the scientific community agreed to accept Hubble's original 1929 paper, and dismiss the doubts in all his later published works.

---

[89] C. Lämmerzahl, O. Preuss and H. Dittus, Is the Physics within the Solar System Really Understood, University of Bremen, Germany, Max Planck Institute for Solar System Research, April 12, 2006, p. 2

This raises a specific concern: whether cosmology is motivated by a desire to preserve a particular view of the cosmos instead of aspiring to practice unfettered inquiry.

The redshift controversy indicates that if a scientist refuses to support the *status quo*, he might be reinterpreted or even ostracized. What is the reason physics classes don't teach Hubble's actual conclusion? Is it the result of prejudice, or the fault of a particular theory?

The significance of Hubble's evidence for rejecting redshift as a Doppler effect can't be underestimated. No person tries to prove himself wrong, what to speak of a respectable leader in his chosen field like Edwin Hubble. The value of his observational data can't be denied due to his pre-eminent position as the greatest astronomer of the twentieth century. But as they silenced Humason with neglect, so they silenced Hubble with respect.

Physics courses continue to promote Hubble's Law, the 1929 explanation, as the direct physical observation of recession speed which is consistent with the solutions of Einstein's equations of general relativity for a homogeneous isotropic expanding universe.

Based on this version, galaxies are receding at great speeds from the observer. That's why we say the space-time volume of the observable universe is expanding.

A "homogeneous isotropic cosmos" means that the universe looks the same to every observer. So every light source will be seen as receding in every direction away from any observer anywhere in the cosmos! Moreover, the properties and laws of the entire cosmos are the same for all observers. This scheme is enshrined in the Cosmological Principle.

The assumption is that Earth does not occupy a special or privileged position in the universe. Science teaches this principle in contra-distinction to the traditional creation story which teaches that Earth does hold a unique position in the universe.

Before court adjourns for the day, our attorney poses two questions:

1. Has redshift been validated by modern scientific data?
2. Has the Big Bang theory been confirmed by redshift validation?

13
Redshift Resistance

An overwhelming array of media people have shown up in court this morning. They reflect the interest this case has generated. One newspaper echoed the phrase "apparent velocity" in its story of the trial.

Outside the justice building a reporter from the local Channel Seven Eyewitness News team brings viewers up to date on the proceedings. When the prosecuting attorney arrives on the scene, he is invited to comment on the case.

In a brief interview on camera, he explains that a great divergence of opinion exists about redshift interpretation. Modern physics courses teach students that all stellar bodies will show recessional speed away from every observer from any location in the universe. Even though the only location from which any human being has ever observed the cosmos is from our tiny solar system.

"Today I intend to sort out the facts from the fantasies," he says before rushing off. Several bystanders snap his photo.

HUBBLE IN TROUBLE

Scientists who promote the Big Bang theory define redshift as the stretching of the wavelengths of starlight due to stars receding at tremendous speeds away from Earth. The recession is caused by the expansion of space. The faster the recession, the greater the wavelength will be stretched. The larger the redshift the further away the star is said to be.

Historically, Herbert Ives demonstrated in 1939 that the bending of starlight near the Sun is a result of the slowing down of light in gravitational fields, not because of a warping of space-time. As a beam of light passes the sun, the part of the beam nearer to the sun will be slowed more than the part of the beam further away. The Sun acts the same as a lens, since lenses slow the speed of light, which we see as refraction.[90]

The longer light travels the more it interacts with everything in space. Could the frequency of a beam of light diminish the longer the beam travels through space?

Hubble wondered if an undiscovered law caused the energy of light to dissipate when it spreads at great distances and interacts with the medium of interstellar space. Using this hypothesis he posited that light's energy loss accounts for the shift to the red end of the spectrum.

> "...light loses energy in proportion to the distance it travels through space. The law, in this form, sounds quite plausible. Inter-nebular space, we believe, cannot be entirely empty. There must be a gravitational field through which the light-quanta travel for many millions of years before they reach the observer, and there may be some interaction between the quanta and the surrounding medium...Light

---
[90] Journal of the Optical Society of America, 29:183-187, 1939

> may lose energy during its journey through space, but if so, we do not yet know how the loss can be explained."[91]

As more data became available, Hubble wrote in 1942 that the redshift interpretation complicated and distorted our understanding of the cosmos. He continued to propose the existence of an unidentified principle of physics.

> "Thus the use of dimming corrections leads to a particular kind of universe, but one which most students are likely to reject as highly improbable. Furthermore, the strange features of this universe are merely the dimming corrections expressed in different terms. Omit the dimming factors, and the oddities vanish. We are left with the simple, even familiar concept of a sensibly infinite universe. All the difficulties are transferred to the interpretation of redshifts which cannot then be the familiar velocity shifts...
>
> "Meanwhile, on the basis of the evidence now available, apparent discrepancies between theory and observation must be recognized. A choice is presented, as once before in the days of Copernicus, between a strangely small, finite universe and a sensibly infinite universe plus a new principle of nature."[92]

Clearly, Hubble thought an unidentified principle of physics was responsible for redshift. In a 1949 article for the *Journal of the American Astronomical Society*, Guy Omer focused on the problems that Hubble observed.

---

[91] *The Observational Approach to Cosmology*, Oxford, Clarendon Press, 1937, p. 30
[92] Edwin Hubble, "The Problem of the Expanding Universe," American Scientist, Vol. 30, No. 2, April 1942, p. 99

> "E. Hubble has shown that the observational data which he has obtained do not agree satisfactorily with the homogeneous relativistic cosmological models...The model has a short time scale. The present age of the model must be less than 1.2×10^9 [1.2 billion] years. This is about one-third the recent estimation of the age of the earth as an independent body, made by A. Holmes. This is probably the most serious difficulty of the homogeneous model. Because of the unrealistic aspects of the homogeneous relativistic model, Hubble proposed an alternate model which would be essentially static and homogeneous in which the red shift would be produced by some unknown but nonrecessional mechanism."[93]

Hubble understood that defining redshift as a Doppler effect leads to a paradox. If redshift is a velocity indicator, the universe becomes too small and too young to accommodate the theory of biological evolution!

> "A universe that has been expanding in this manner would be so extraordinarily young, the time-interval since the expansion began would be so brief, that suspicions are at once aroused concerning either the interpretation of redshifts as velocity shifts or the cosmological theory in its present form."[94]

The Doppler interpretation disturbed Hubble even more because he couldn't disprove it. From his quote it seems that he's offering two options:

---

[93] Guy C. Omer, Jr., "A Nonhomogeneous Cosmological Model," Journal of the American Astronomical Society, 109, 1949, pp. 164-165
[94] *The Observational Approach to Cosmology*, Oxford, Clarendon Press, 1937, p. 46

1. reinterpreting the redshift as being velocity shifts,
2. reinterpreting the present cosmological theory.

The cosmological theory he's referring to posits the universe is curved as opposed to the flat universe picture (with no overall warping of space-time) suggested by current experiments which we discussed in Chapter Ten. This leads one to believe that his troubles would be resolved by alternative (2).

Hubble's troubles were not over yet, however. For the rest of his life, he searched for a viable explanation of redshift. It was not only a dilemma for him, but also for the modern creation story. In his 1937 book he was candid with his doubts about the 1929 redshift interpretation.

None of his later works represent his original paper of 1929. Right up to his death in 1953, he confided to Nobel Prize winner Robert Millikan that redshift should not be interpreted as a Doppler effect. This admission lent strong support for Fred Hoyle's Steady State universe that became a popular theory in the 1960s.

Although Hubble rejected his original explanation, the astronomical community chose to ignore his later published works and only accept his initial 1929 paper. What does that mean?

It means that the scientific establishment that never sat for decades in front of the world's most powerful telescope decided to overrule the astronomer who did. That's how the explanation for the expanding universe theory became Hubble's 1929 paper on the redshift of starlight.

Never mind that Hubble didn't accept his own "law" or that redshift interpretation was an arbitrary choice. Apparently, those were minor details.

Seeing Red

A hundred years ago science taught that light travels through a pristine environment in space and makes its own electromagnetic medium as it travels. General Relativity theory posits that space is a vacuum. As light travels from a receding star no physical substance exists in space which can interact with it. Einstein described this as "c in vacuo".

Based on this idea, the explanation for redshift in starlight might be due to the motion of the star, meaning the specific recession of the star away from Earth. This was the foundation for the expanding universe theory.

In the 21st century we know that space is not empty; nor is it a perfect vacuum. Edwin Hubble wondered if the frequency of a light beam would lose energy the longer it traveled through space. He couldn't explain how it worked but his experiments had proven to him that redshift could not possibly be a Doppler shift of velocity. The scientific community, however, was not convinced.

Today, the calculation of a star's recession speed is still known as Hubble's Law. But science needs data. Otherwise it's only conjecture. Why is it so important for cosmology to maintain Hubble's Law?

The answer is simple. If we interpret redshift as a Doppler shift of velocity for expanding galaxies, then we can calculate how long the expansion has been going on. By running the figures in reverse we can determine when the universe began—the Big Bang. Therefore, "Hubble's Law" is needed to calculate the age of the universe and justify the Big Bang model, even though Hubble later refuted his own hypothesis.

Is there new data to substantiate redshift as a true Doppler effect in the 21st century? Not according to some physicists who have revisited the problem. They suggest various alternative explanations.

In his 1990 article, "Implications of the Compton Effect Interpretation of the Redshift", John Kierien presents that redshift is caused by the Compton Effect, not the Doppler Effect. The Compton Effect is the result of high-energy photons colliding with an atom or molecule. The scattered radiation causes a wavelength shift that can't be explained by classical wave theory.

Alternatively, Tom Van Flandern suggests that redshift is caused by friction between light waves and the classical graviton medium it travels through.

Astronomer Halton Arp, a former associate of Edwin Hubble, proposed another explanation based on direct observation. He discovered that the biggest redshifts belonged to quasars — brilliant, point-like objects that are supposedly at the edge of the universe. Yet some quasars which were rather close to each other had vastly different redshifts. Moreover, younger quasars had higher redshifts.

For 29 years Dr. Arp was a staff astronomer at Hale Observatories, which included the Mount Wilson and Mount Palomar observatories. He often found high redshift quasars in the middle of low redshift galaxies which indicated to him that velocity was not the governing factor.

In a paper published in 1967, he wrote: "It is with reluctance that I come to the conclusion that the redshifts of some extragalactic objects are not due entirely to velocity causes."

Surprisingly, quasars were often suspiciously close in the sky to relatively nearby spiral galaxies. Based on his observational data, Arp posited that quasars were not so distant after all and a red shift was not always an indication of distance. Perhaps redshift could be caused by some unknown property of physics.

In his 1998 book, *Seeing Red*, Arp wrote: "if the cause of these redshifts is misunderstood, then distances can be wrong by factors of 10 to 100, and luminosities and masses will be wrong by factors up to 10,000. We would have a totally erroneous picture of extragalactic space, and be faced with one of the most embarrassing boondoggles in our intellectual history."[95]

In other words, the picture of cosmic evolution given by the Big Bang model, whereby the universe slowly condensed into stars, galaxies and creatures over almost 14 billion years would have to be abandoned.

Despite this "most embarrassing" effect of not getting it right, various explanations for redshift abound nowadays. Arp eventually proposed that redshift was an indicator of age because he observed newer objects with higher redshifts.

Physicists who favor interpretations based on mathematical formulations tend to ignore some observational data, consigning them to the anomalies category. There is no data to support popular redshift explanations, yet the data that indicates otherwise is neglected.

When should we place facts before protocol? Our actions are determined by our beliefs. Our beliefs are shaped by the teachings we hear from those authoritative sources we have come to respect. What happens when doctrines from esteemed authoritative sources are called into question?

Like Humason before him, Arp was ostracized from astrophysical circles and his telescope viewing schedule was revoked. He was banned because his extensive research on the redshifts of quasars and galaxies showed that redshift should not be used to prove the universe was expanding.

Regardless of whether Arp's explanation of redshift is correct, we want to know whether genuine scientific inquiry is still the accepted norm once it strays beyond the prevailing assumption?

The marginalization of Arp, and Humason before him, seems to indicate the astronomical establishment's reluctance to consider particular data. This aspect of cosmology inhibits

---

[95] Halton Arp, Seeing Red: *Redshifts*, Cosmology and Academic Science, p. 1

and restricts free inquiry, which is the life blood of the scientific endeavor. One may question whether this is due to a biased orientation to preserve the *status quo*.

Apparently, to kill the sacred cow one becomes an infidel. Thus, science is inhibited by this attitude.

Regarding Hubble's hypothesis that light loses energy over time and distance, Arp didn't agree because his observational data showed no increase in redshift from light traveling through dense galactic material.

Physicists Charles Misner, Kip Thorne, and John Wheeler, concur. The following explanation regarding the "tired light" concept is from their well-known book *Gravitation*.

> "If the energy loss is caused by an interaction with the intergalactic matter, it is accompanied by a transfer of momentum; that is, there is a change of the direction of motion of the photon. There would then be a smearing out of images; a distant star would be seen as a disc, not a point, and that is not what is observed...if the decay of photons is possible at all, those in radio waves must decay especially rapidly!
>
> "There is no experimental indication of such effects: the radio-frequency radiation from distant sources is transmitted to us not a bit more poorly than visible light, and the redshift measured in different parts of the spectrum is exactly the same..."[96]

---

[96] Misner, Thorne and Wheeler, *Gravitation*, New York, 1973, 25th printing, p. 775

MORE DOPPLER EFFECT PROBLEMS

Although the Doppler interpretation of redshift was widely accepted, it caused a major "side effect" problem of major proportions. It meant that the outer galaxies were receding from Earth faster than the speed of light.

General Relativity now faced a contradiction to Einstein's most precious idea—the inviolability of the speed of light.

In Chapter Nine we discussed the following:

1. gravitational force may exceed light speed
2. cosmic inflation exceeds the speed of light
3. quantum particle communication exceeds light speed

How did cosmologists deal with the fourth anomaly that defied the speed of light? They simply changed the "expanding" universe into the "exploding" universe. The key component was that the explosion created new space faster than light can travel. This explanation proposes that galaxies and stars are being pulled along with the explosion so it only appears they travel faster than light. That's how the Big Bang model was validated.

Is this new space being created faster than the speed of light in all directions simultaneously? That's unimaginably fast! Yet, the mechanism that continually creates space in all directions at an unfathomable speed is unknown. We must simply accept the explanation, dark energy, on faith.

That primeval singularity which contained the entire substance of the future universe, and which was infinitesimally smaller than the period at the end of this sentence, decided for some unknown reason, or at least unknown to science, to explode. We are still embedded in that continuing explosion which produces everything we see in the universe.

Stephen Hawking already informed us in Chapter Seven that the Big Bang theory cannot, and does not, provide an explanation for an initial condition. It only mathematically describes the general evolution of the universe going forward after that first event. Physicists now admit Einstein's equations break down before the point of the expansion of the singularity (which contained infinite curvature and density).

Even examining this post-event period one might ask, "Into what medium is the universe expanding?" Of course, this begs the question, "In what medium did the singularity exist before space was created?"

Most physicists will say the singularity is spacetime itself so nothing exists outside. Thus, they repudiate these as "meaningless questions."[97]

If they have no answer they could just say, "We don't know right now." That's honest. Branding questions as meaningless is the kind of retort we get from priests and mullahs, which is why people lose faith in religion.

If the universe is infinite it would go on forever and have no boundary, meaning it doesn't have anything to expand into because nothing outside infinity would exist. Nor can more stuff be added to infinity to make a greater infinity. So the answer from a philosophical perspective is that the question isn't relevant when dealing with infinity.

Is there evidence to support an infinite universe? Not yet. It's an assumption of cosmology that we accept on faith.

Alternatively, consider a finite universe with a boundary. In that case it is legitimate to claim we could be expanding into 'something'. However, the boundary is so distant and is expanding faster than light speed, so that 'something' could never be detected. The expansion assures us that a boundary would always lie beyond our detection. Moreover, we are stuck within the space of our cosmos with no way to observe anything outside our space. So even if it was finite it's impossible to know

---

[97] Misner, Thorne and Wheeler, *Gravitation*, W. H. Freeman and Co, 1973, 25th edition, p. 739

what the universe is expanding into.

However, to question how the singularity came into existence, and how all the matter and energy in existence was crushed into a subatomic size, or how the fine-tuned laws came into existence, should be the quest of scientific minds. Another question is this: what happened at the absolute beginning to create space if literally nothing existed previously?

Science has no answers for these problems, but to deny questions reeks of a dogmatic knee-jerk reaction. We must ask questions in order to find answers. In the 21st century, general relativity theory is experiencing a crisis. Stephen Hawking has already revealed the painful truth in two of his earlier books.

> "We already know that general relativity must be altered. By predicting points of infinite density—singularities—classical general relativity predicts its own downfall... When a theory predicts singularities such as infinite density and curvature, it is a sign that the theory must somehow be modified."[98]

Hubble's assertion that redshift is not a Doppler Effect remained an enigma for astronomy. But how is redshift interpreted in the 21st century?

The modern interpretation is divided into several ideas. When a *nearby* body is moving away from the earth, in that specific case science retains the original interpretation of redshift as a Doppler Effect. In such cases the effects of the expansion of space are negligible.

When dealing with *distant* objects the expansion of the universe becomes an important factor. According to general relativity, the expansion is not due to objects moving away from each other. Rather, it's the constant expansion of space be-

---

[98] Stephen Hawking, *Black Holes and Baby Universes*, New York, Bantam Books, 1994, p. 92. *A Briefer History of Time*, New York, Bantam Dell, 2005, pp. 102, 84

tween objects that stretches. Light moving through stretched space will also become stretched. This causes its wavelength to shift to the red end of the spectrum. This redshift is based on Einstein's equations and is called a "cosmological redshift".

In other words, a cosmological redshift is a measure of the expansion that the universe has undergone from the time the light was emitted and the time it was received. Measurements of this redshift allows us to estimate distances to other galaxies

There is a third phenomenon known as a "gravitational redshift". This idea describes how gravity's effect on space-time changes the wavelength of light moving through that space-time. The modern explanation for Humason's "apparent velocity" is defined as an increase in distance that occurs due to the stretched condition of expanding space.

So in terms of redshift explanation, where does that leave us now?

Previously, physicists claimed redshift was a Doppler Effect to establish an expanding universe. Now they claim cosmological redshift is due to the expansion of space itself and is not a Doppler velocity effect. What has changed? They simply switched the cart and the horse.

What do these disclosures indicate? Is it feasible that the Big Bang didn't happen?

No Big Bang

In 1991 Physicist Eric Lerner published his major work on the origin of the universe, *The Big Bang Never Happened*.

"If the Big Bang hypothesis is wrong," Lerner wrote, "then the foundation of modern particle physics collapses and entirely new approaches are required. Indeed, particle physics also suffers from an increasing contradiction between theory and experiment."[99]

The arguments in Lerner's book attracted a new breed of physicists. Thomas Van Flandern sums up the anomalies and contradictions in the Big Bang theory with his book *Dark Matter, Missing Planets, and New Comets*.

> "The Big Bang theory is the accepted model for the origin of the universe. This theory requires us to accept the following...that all the matter and energy in the entire universe were contained in an infinitesimal point at the 'beginning'; that for some unknown reason it all exploded; that space and time themselves expanded out of that explosion; that at first space expanded faster than the speed of light; that the explosion was so uniform it emitted an almost

---

[99] Eric J. Lerner, *The Big Bang Never Happened*, New York, Random House, 1991, p. 4

perfectly uniform radiation everywhere; and the same explosion was non-uniform enough to create the observed, quite irregular matter distribution in the universe; that the chaos from the explosion eventually organized itself into the structures presently seen in the universe, contrary to the principle of entropy (which basically states that you shouldn't get order out of chaos); that all matter in the universe expands away from all other matter as space itself continues to expand, although there is no center; that the expansion of space itself occurs between all galactic clusters and larger structures, but does not occur at all on scales as small as individual galaxies or the solar system; that vast assemblies of galaxies stream through space together relative to other assemblies; and that immense voids separate immense walls of galaxies, all condensed from the same explosion."[100]

Every one of Van Flandern's eleven points is an unresolved anomaly with the Big Bang model. In other words, the major tenets of the model have never been established as factual.

In his book *Cult of the Big Bang*, William Mitchell provides extensive analysis for over 30 significant problems with the Big Bang hypothesis. Normally, any model with so many problems would be discarded and the search for an alternative would ensue. When so many doubts are raised we may question why the astrophysics community continues to minimize the anomalies?

Mitchell's reply is that enormous effort has already been spent to support the Big Bang model. Still, he questions why

---

[100] Tom van Flandern, *Dark Matter, Missing Planets and New Comets*, revised edition, Berkeley, CA: North Atlantic Books, 1993, p. xvi

talented people take part in minimizing faults. He concludes there must be other factors involved to degrade the process.

Physicists are considered "dissidents" within the cosmological establishment if they no longer accept the Big Bang model. But anomalies need to be resolved, not swept under the rug to uphold a struggling theory.

Even back in 1969 when the Big Bang theory was in its infancy the respected astronomer Robert Dicke published a sobering assessment that it was an unlikely event.

> "The puzzle here is the following: how did the initial explosion become started with such precision, the outward radial motion became so finely adjusted as to enable the various parts of the Universe to fly apart while continuously slowing in the rate of expansion? There seems to be no fundamental theoretical reason for such a fine balance. If the fireball had expanded only 0.1 per cent faster, the present rate of expansion would have been 3×10^3 times as great. Had the initial expansion rate been 0.1 per cent less and the Universe would have expanded to only 3×10^{-6} of its present radius before collapsing. At this maximum radius the density of ordinary matter would have been 10^{-12} gm/cm^3, over 10^{16} times as great as the present mass density. No stars could have formed in such a Universe, for it would not have existed long enough to form stars."[101]

Professors can teach convincingly that the Big Bang is real. Yet anomalies and contradictions offered by so-called "dissident" cosmologists beg for answers. Van Flandern's critique of Big

---

[101] Robert H. Dicke, "Gravitation and the Universe," Jayne Lectures for 1969, *American Philosophical Society*, Independence Square, Philadelphia, 1970, p. 62

Bang theory reveals that he doesn't buy the foundational premises of the Big Bang model.

The major limitation of the Big Bang theory is the singularity itself. Even the near uniform temperature of the cosmos (approx 3º Kelvin) means that the 13.8 billion year age of the cosmos does not allow enough time to reach temperature equilibrium, based on our present understanding, even if inflation ensued faster than light speed after the explosion from the singularity.

The Big Bang theory is definitely in a crisis condition. On page 20 of the May 22, 2004, edition of *New Scientist*, Eric Lerner presented "An Open Letter to the Scientific Community." The letter was signed by thirty-three scientists as representing their view.

In this open letter Lerner conveyed that "…the Big Bang is not the only framework available for understanding the history of the universe. Plasma cosmology and the steady-state model both hypothesize an evolving universe without beginning or end."

In 2005, *New Scientist* offered a follow-up article "The End of the Beginning" by Marcus Chown. The article quotes Eric Lerner regarding the Big Bang. "This isn't science. Big Bang predictions are consistently wrong and are being fixed after the event."

Lerner is referring to the concept of retrodiction (the opposite of prediction) which takes place after the discovery of new observations. Yet some physicists still say the theory predicted the outcome although no advance notice before the discovery was foretold.

In terms of erroneous predictions the magazine's editor added the following: "So much so, that today's 'Standard Model' of cosmology has become an ugly mishmash comprising the basic Big Bang theory, inflation, and a generous helping of dark matter and dark energy."[102]

The Standard Model [a theory of forces and interaction of particles to explain how the universe works] was the unification of elementary point particles for which Steven Weinberg received the Nobel Prize. But there were glaring omissions; gravity was not included, for example.

Despite the shortcomings of the Standard Model, Wikipedia tries to explain Big Bang cosmology in glowing terms to the general public:

> "The Big Bang is a well-tested scientific theory which is widely accepted within the scientific community because it is the most accurate and comprehensive explanation for the full range of phenomena astronomers observe. Since its conception, abundant evidence has arisen to further validate the model."

Nevertheless, today we find a growing cadre of cosmologists who no longer accept the Big Bang theory, preferring a model where time and space had no beginning; a timeless infinite universe.

Not only is Eric Lerner pushing for this, other so-called dissident cosmologists like Halton Arp, Michael Ibison, Hermann Bondi, Paul Marmet, Jayant Narlikar, Sisir Roy, William Mitchell, and Tom Van Flandern, are among many who do not accept the Big Bang origin of the universe. They prefer, as a more viable alternative, the Steady State theory originally proposed by Fred Hoyle, and accepted by Arthur Eddington among other reputable 20th century astronomers.

Even Stephen Hawking no longer accepts the Big Bang model as viable. That's why he proposed his new alternative theory, dubbed the Grand Design, based on M-theory. Overall, the Big Bang model is misleading and even doubtful consider-

---

[102] Marcus Chown, "The End of the Beginning," *New Scientist*, July 2, 2005, p. 30

ing the many anomalies cited above.

Contrary to what is being presented to the gullible public, a battle is now raging within astrophysical circles. Two ideas might save the Big Bang theory: dark matter and dark energy. Is a resolution possible?

At this point the judge decides the jury has heard enough for one day, and calls for an end to the proceedings. "We'll carry on from here tomorrow," he says with a big bang of his gavel.

14

Cosmology in Conflict

The discussion in court this morning focuses on the two outstanding theories of 20th century physics—General Relativity and Quantum Mechanics.

Basically, relativity theory deals with massive things and quantum theory deals with microscopic things. Our present understanding is that matter functions differently according to size. We also know it behaves in a different way according to temperature; like water, ice, and steam.

Our attorney walks toward the jury and emphatically states, "Let's see what other conditions prompts matter to act differently."

Physicists are struggling to come to terms with the origin of the "bang" in the Big Bang. It's what caused the expansion of time and space. Evidently, it was subatomic in size yet had enormous mass. It also had, supposedly, infinite temperature, density, and curvature. Physicists call it a "singularity." We are told it was the cause of the Big Bang billions of years ago.

Cosmology needs both relativity theory and quantum theory in order to understand this singularity. The problem is these theories do not agree with each other. They're like oil and water—they don't blend. The basis of each is totally different.

Still, physicists are looking for an explanation that combines both theories. They believe there is one master formula that will explain everything both infinite and infinitesimal. They want a grand unified theory, a GUT. But they are faced with a huge problem. They have no ability to test it.

Science lacks the ability to reproduce a black hole or a singularity in the laboratory. Yet, a singularity is widely accepted to be a black hole, and there are many in our universe. So there should be some evidence that they are spawning off into universes, causing space-time to explode. That evidence is missing, which leaves us with no observational data to support the theory. How will we know if a GUT actually represents reality?

Dr. Paul Frances, physicist at the Australian National University, puts this problem in perspective. "There are so many rival contenders for a GUT, but basically we don't have the faintest idea. When we extrapolate well beyond what can be tested we're asking for trouble."

The trouble he refers to is the following: "When there's no data to back up theories it just becomes like a beauty contest. It's not science and it will remain not science."

We already discovered that mathematics doesn't always represent the real world. That would include the equations which purport to explain the birth of the real world. Science has always believed in 'what you see is what you get.' A GUT is only a mathematical academic explanation unless and until it becomes a verifiable explanation.

We observe that the universe comprises stars, galaxies, and planets. On the other hand, no scientist has detected dark matter or dark energy. Yet they must be there for the Big Bang theory to work. I don't think speculation about undetectable matter and energy is a strong basis to confirm a theory. If tools disappear from my garden, who will take me seriously if I say it's unknown elves and fairies?

Dark Matter

Originally, the theory of dark matter arose by noticing discrepancies between the mass of large astronomical objects (determined from their gravitational effects) and the mass calculated from the "luminous matter" they contain. It wasn't invented to sustain the Big Bang theory directly, though it became useful for that. Let's examine how this came about.

Physics tells us that when objects revolve around a center they want to escape from a circular motion. This is due to tangential velocity. There must also be a force that pulls the object towards the center to keep it from flying off into space.

Stars and gases are held in place by the gravity of the galaxy. Astrophysicists can calculate how fast stars and gas clouds are moving. Then they calculate how much gravity is needed to keep them in orbit.

The calculations revealed a problem. Galaxies weighed ten to a hundred times more than they should, according to our theories about gravity. To keep the stars and gases orbiting the galaxy, ten to a hundred times more gravity would be necessary. That's how the dark matter paradox began.

In all the galaxies under examination, unseen matter had to be there to provide enough gravity to hold everything in orbit. There had to be something in the galaxies that they couldn't see (dark) which had mass (matter) and that matter had gravity.

Thus, dark matter was hypothesized to solve a problem—to account for the missing mass. So the existence and properties of dark matter were posited by unknown gravitational effects on visible matter.

One thing was obvious. This dark matter was not opaque because it would block our sight, so it must be transparent. Scientists figured out one other thing. If dark matter was made of protons and neutrons, like normal matter, it should have been involved in nuclear reactions and thus produced different elements as a result of the Big Bang. But we don't see any indication of the existence of strange elements. Therefore, scientists concluded that dark matter is not made of neutrons and protons.

So what is dark matter made of? The most widely accepted explanation is that it might be composed of weakly interacting massive particles (WIMPs) that interact only through gravity and the weak force. Detectors are looking out for WIMPs, but not one has been stumbled upon.

Alternatively, dark matter might be made of some unknown subatomic particle. A search for this particle is a major effort in physics. Astronomers began to map the motions of hundreds of stars in the Milky Way to figure out how much dark matter must be tugging on them from the vicinity of our sun. But they came to a surprising conclusion—there was no dark matter to be found.

The stellar motion implied that all the stars within 13,000 light-years of Earth were gravitationally attracted by the visible material and not by some invisible material.

If the analysis of the data is verified it indicates that dark matter can't be detected in our space within the Milky Way. Then its distribution must be quite unusual to avoid our region in space. This could be a serious blow to the burgeoning dark matter hypothesis.

Yet, astrophysicists believe that dark matter is the most prevalent kind of matter in the cosmos. They do believe it affects the movement of galaxies, so astronomers began using high-energy telescopes to study the gravitational force of galaxy clusters—vast flocks of galaxies bound together by gravity. There is optimism that observations of galaxy clusters can provide clues about dark matter.

Using space-borne telescopes from NASA and ESA (European Space Agency) astronomers agree that 80 percent of galaxy clusters are filled with dark matter, while hot gases and the galaxies themselves occupy the rest of the intervening space.

Hydrogen is the most prevalent hot gas, which emits x-rays that are visible via telescopes.

Additional x-rays from subatomic particles with distinctive wavelengths were also detectable by telescope. Surprisingly, an unknown x-ray wavelength was seen that had not been noticed previously. Hence, theorists searched for an explanation beyond the realm of known matter.

A theory was proposed that the abnormal x-ray wavelength might be due to the deterioration of a subatomic particle called a sterile neutrino. Not surprisingly, this unique particle has also never been detected by astronomers, yet they believe it interacts with matter via gravity. This belief is one way to resolve the problem.

In the meantime, what do physicists *know* about dark matter? Well, there's a lot of it and it doesn't shine. It's not made of normal protons and neutrons. It's very heavy and it's transparent. That's the sum total of their understanding of dark matter.

They believe it affects the movement of galaxies but no one knows how far out it goes. Yet, the further out they measure the more there seems to be. Unfortunately, we run out of orbiting stars to accurately measure how far off it's moving. Science can only measure as far as the cosmological horizon, and/or the CMB.

There are two ways to understand cosmology; by observation or by theory. Physicists define dark matter as a massive "something" that's invisible and undetectable. But it does exactly what they require it to do—account for matter that seems to be missing. Of course, the massive amount of undetectable dark matter, or more accurately, the missing gravity to maintain the galaxies in balance, could just as easily be caused by an undiscovered principle of physics.

Another problem: whatever dark matter is, where is it located? Some theorists thought it might originate from the giant black hole lurking at the center of our Milky Way galaxy. Yet physicists have now ascertained that this dark matter seems to pervade the whole universe, as far as they can tell. So whatever it is they know it's more spread out than the stars.

Another reason why physicists posit the existence of dark matter is that objects at the outskirts of galaxies move way too fast according to our 21st century understanding of gravity. They should be moving slower to remain in orbit according to the amount of gravity generated by visible matter.

Of course, if our understanding of gravity is incorrect there may be no need to posit dark matter. Perhaps there is a deviation from the inverse square law at galactic distances.

We already learned that science still hasn't got gravitational force sorted out. We accept Newton's laws of gravity because they work on the scale of our solar system. But is there an undiscovered law that only works on a massive scale?

A new theory called Modified Newtonian Dynamics, or MOND, says that gravity doesn't obey the laws we think it should. It obeys an undiscovered law. In other words, a "hidden variable theory" postulated by both Einstein and Hubble.

Dr. Paul Davies explains the conundrum cosmology now found itself in:

> "While string theorists were frantically developing their abstract models, astronomers made a series of discoveries that exploded like a grenade in the entire theoretical physics paradigm, throwing string theory as well as cosmology into turmoil."[103]

It was the discovery in 1998 that the expansion of the universe was accelerating. Prior to this, every physicist assumed it was slowing down.

TED Talks is a daily video podcast of leading thinkers and doers who give brief updates on their work. On a Febru-

---
[103] Paul Davies, *The Goldilocks Enigma*, Penquin Books, London, 2007, p.130

ary/2012 podcast, theorist Brian Greene put this discovery in perspective.

> "Here's the surprise: They found that the expansion is not slowing down. Instead they found that it's speeding up, going faster and faster. That's like tossing an apple upward and it goes up faster and faster. Now if you saw an apple do that, you'd want to know why."

The result was a new hypothesis called dark energy. What is it? Physicists haven't the foggiest idea.

Dark Energy

The dark matter conundrum was not good for physics. But then scientists made an even worse discovery—dark energy. It turns out that the universe is mostly mysterious stuff which nobody can see.

What we *can* see comprises only 4.0 percent of the universe. That's all we can know about the cosmos. Out of this, a mere 0.4 percent is all the visible stars and galaxies. The 3.6 percent that's made of normal matter and atoms is probably intergalactic gas. Nobody knows for certain. Dark matter constitutes 22 percent of the universe. Dark energy makes up 74 percent of the entire cosmos.

According to the laws of physics we would expect that after the Big Bang, the initial rapid expansion would start slowing down as time passed. Gravitational force is attractive, so gravity should slow down the cosmic expansion and even bring the acceleration to a halt. Then, the gravitational force should start collapsing everything back together again. This is the normal physical process based on what we know about gravity.

But our observations tell us something different. They tell

us that the universe is accelerating at an increasing rate. It's not what we expected. It means that either our understanding of gravity is wrong, or there is something else out there we don't know about—a missing element.

Gravity is a pull force. If everything is accelerating it means there must be a push force. That's how physicists came to the conclusion that something else is out there. Of course, a force that repels other things didn't exist in physics so they had to invent one. Since they couldn't detect anything, they called it dark energy.

We know that normal forces are stronger when they're closer together and weaker when they are farther away. Gravity becomes weaker as things move further apart. The Inverse Square Law describes how fast a force F becomes weak with the distance r ($F \sim 1/r^2$). There are other forces that also become weaker with distance. Some do it faster, some do it slower, but no force gets stronger with distance.

Therefore, cosmologists had to formulate a force (dark energy) which gets stronger and pushes away at greater velocity the further away things are. They had no choice but to invent a force, which had to be out there, in order to account for their present understanding of gravity.

So what do we understand about dark energy? Well, it's a mysterious force that makes up 74 percent of our universe. It pushes rather than pulls. It gets stronger when things are farther away, and we don't know what causes it. This is a strange force, indeed.

In spite of all this, Stephen Hawking assures us, "The scientific account is complete." Is he telling the truth, the whole truth, and nothing but the truth?

Astrophysicists continue to search for a dark energy explanation. There are as many versions as theorists working in the field. A possible contender for that explanation is Einstein's cosmological constant. But that idea was entirely different. Einstein's force acted to keep the universe static, not to push it away.

Still, what exactly is the cosmological constant? It's not some electromagnetic force, neither a strong or weak nuclear force, nor is it the force of gravity. Actually it is just a conceived mathematical force that has no basis in reality and is not considered by any other scientists except cosmologists.

Brian Greene puts his own spin on the issue. "Well, according to Einstein's math, if space is uniformly filled with an invisible energy, sort of like a uniform invisible mist, then the gravity generated by that mist would be repulsive, repulsive gravity, which is just what we need to explain the observations."

So an invisible energy in space conveniently becomes repulsive gravity. Greene assures us that "this explanation represents great progress." But it's based on the word "if" and on conceived mathematics.

GALAXY FORMATION

The anomalies of modern cosmology are slowly coming to the surface. Even the best kept secrets are being revealed. We don't know the cause of gravity. We find out that 96 percent of the universe is undetectable and remains a mystery.

The formation of galaxies is another enigma. Even back in 1975, James Binney wrote:

> "The real problems of galaxy formation remain very much unsolved. The greatest difficulty is that we still have no idea what induced the formation of the first bound objects in an expanding universe."[104]

Ivan R. King, former professor of astronomy at the University

---

[104] *Nature*, 255:275-276, 1975; See also: J. Binney, 1981b, in The Structure and Evolution of Normal Galaxies, ed. S. M. Fall and D. Lynden-Bell, Cambridge: Cambridge Univ. Press. J. Binney, 1982b, *Annual Review of Astronomy and Astrophysics*, 20, 399

of California, Berkeley, declared the galaxy formation problem to be a "flagrant scandal that is rarely mentioned in public."[105]

A 2004 press release by Karl Glazebrook publicized the following admission from a study by Johns Hopkins University:

> "It seems that an unexpectedly large fraction of stars in big galaxies were already in place early in the universe's formation, and that challenges what we've believed. We thought massive galaxies came much later...
>
> "This was the most comprehensive survey ever done covering the bulk of the galaxies that represent conditions in the early universe. We expected to find basically zero massive galaxies beyond about 9 billion years ago, because theoretical models predict that massive galaxies form last. Instead, we found highly developed galaxies that just shouldn't have been there, but are."[106]

The most distant galaxies are apparently 10 billion light years from us. We would expect the formation of such distant galaxies to look 10 billion years younger since it takes 10 billion years for their light to reach us. After all, we are seeing them as they were billions of years in the past. Yet they look the same as the galaxies in our own neighborhood.[107]

The fact that they look the same creates an intractable situation for astrophysicists. How to explain the discrepancy? Some physicists posit that galaxies mature rapidly and level off millions of years later. Of course, there is no data to verify this explanation. Furthermore, it's in conflict with current explanations of galaxy formation.

But the problem is even deeper than one might expect. A

---

[105] *The Evolution of Galaxies and Stellar Populations*, ed. B. M. Tinsley and R. B. Larson, New Haven: Yale University Observatory, 1977
[106] *Sky and Telescope*, "Old Galaxies in the Young Universe", January 6, 2004
[107] "Most Distant Galaxies Surprisingly Mature", *Science News*, 119:148, 1981

common complaint raised by Steady State adherents is that distant galaxies not only look similar, but various groups of galaxies are so massive it would be impossible for such enormous objects to have formed within the Big Bang time frame.

In an article for *Sky and Telescope* magazine titled "Old Galaxies in the Young Universe," Alan M. MacRobert confirms the dilemma:

> "Astronomers thought they had a nice, clear picture of how galaxies formed billions of years ago—but now the picture is suddenly turning muddy. A team studying the faintest galaxies ever to have their spectra taken is finding far too many big, mature galaxies similar to our Milky Way much too early in cosmic history."[108]

Theoretical physicists are coming to the point of panic. The BBC documentary "Hubble's Deepest Shot is a Puzzle" reports that of 800 exposures in a patch of Hubble's Ultra Deep Field there were far fewer stars being born in the remote past. Apparently, current theories of cosmic evolution are more questionable than certain.

On September 23, 2004, BBC News reported that Dr. Andrew Bunker, the leader of the team studying the faintest galaxies, admitted the following:

> "Our results based on the Ultra Deep Field are very intriguing and quite a puzzle. They're certainly not what I expected, nor what most of the theorists in astrophysics expected."

> "Another exciting possibility is that physics was very different in the early Universe; our understanding of the recipe stars obey when they form is flawed."

---

[108] "Old Galaxies in the Young Universe", *Sky and Telescope*, January 6, 2004

Cosmology in Conflict

Of course, "flawed" means the theory doesn't predict the correct result.

When astronomers discovered the Great Galactic Wall, yet another anomaly was revealed. This wall is a mass of galaxies 500 million light-years by 300 million light-years by 15 million light-years in total area.

In 1989, *Science* magazine reported that such a structure could never have formed in the 15 billion years then assigned to the age of the universe.[109]

The only possible way the Great Galactic Wall could have formed within 15 billion years would be if it had about 100 times the mass it actually has.

Today the universe is considered 13.8 billion years old. In other words, 1.2 billion years younger than thought in 1989. So with a smaller time frame, the calculations just don't add up. These revelations motivated Stephen Hawking to comment: "Either we have failed to see 99 percent of the universe, or we are wrong about how the universe began."[110]

This admission by Hawking was magnified by the discovery of 13 more "Great Walls" of galaxies since his 1989 comment.[111]

Furthermore, the observable scientific data establishes that all distant galaxies appear to be in pristine condition, like our Milky Way. But the known laws of physics preclude pristine galaxies from existing for billions of years because the outer arms in spiral galaxies rotate slower than the dense cores. This is well documented.

Consequently, we should see the arms in older galaxies twisted, or wrapped around and fused into the core. They shouldn't look the same as younger galaxies. Yet in their pristine condition it looks like all galaxies must have formed at the

---

[109] *Science*, November 17, 1989, Margaret J. Geller and John P. Huchra, Harvard-Smithsonian Center for Astrophysics

[110] *Science*, November 17, 1989, p. 11-12

[111] *Astronomy*, "A Cross-Section of the Universe", November 1989; "Southern Super Cluster Traced Across the Sky", January, 1990; "Sky Survey Reveals Regularly Spaced Galaxies", June 1990; Sky and *Telescope*, "The Great Wall", January 1990; "A Universe of Bubbles and Voids", Sept. 1990

same time. Such a conclusion, however, is unacceptable to cosmologists.

According to a report by astronomer Gerardus Bouw, his research does *not* confirm a universe with galaxies that are billions of years old.

> "Evolutionary models have never been successful in accounting for the formation of a single star, let alone a whole galaxy or even a cluster of galaxies (Jones, B. J. T., 1976, Review of Modern Physics, 48:107). Virtually every model in vogue today, which attempts to account for such objects, assumes that they were formed from the collapse of certain density irregularities postulated to be present in the early stages of the Big Bang.
>
> "Without such an assumption, the physics of collapsing gas clouds would not allow for the formation of objects even remotely resembling the major constituents of the universe. A number of explanations have been proposed to account for such density irregularities, including magnetohydrodynamical 'pinch' effects (Fennelly, A. J., 1980, Physical Review Letters, 44:955), but the existence of the required cosmic magnetic field is in doubt and the 3-degree Kelvin blackbody radiation reveals no evidence for any significant clumps of matter at the time believed to be about a million years into the evolution of the Big Bang."[112]

---

[112] *Physics of the Galaxy and Interstellar Matter*, Springer-Verlag, 1987, pp. 352-413. *In the Beginning*, Walt Brown, pp. 23, 30

Shocking New Discovery

Science Daily reports that a team of astronomers claim to have found the largest structure in the Universe: a large quasar group (LQG) four billion light years from end to end. NBC News described it as "mind-boggling" and "theory says it shouldn't exist."

Another headline shouted: "Gigantic quasar cluster shatters long-held theories of space." The study was published January 11, 2013, in the journal *Monthly Notices of the Royal Astronomical Society*.

Astronomer Roger Clowes from the University of Central Lancashire in England stated: "While it is difficult to fathom the scale of this LQG, we can say quite definitely it is the largest structure ever seen in the entire universe. This is hugely exciting, not least because it runs counter to our current understanding of the scale of the universe."

Using data from the Sloan Digital Sky Survey, the international team of researchers discovered the record-breaking cluster of quasars—young active galaxies—stretching four billion light-years across.

The figures are based on the redshift explanation as a velocity shift. The LQG size comes from the assumption that its distance is proportional to its redshift, which is 1.27; the magnitudes of the quasars in the cluster range from 18 to 19.

What does this discovery indicate? Well, researchers claim that the newly discovered LQG is so enormous that our theories predict it shouldn't even exist. The structure is larger than cosmological theory says is possible.

Structures created on the basis of gravity can't be larger than the time it takes gravity to act from one end of the structure to the other, assuming gravity doesn't travel faster than light speed.

Some cosmologists suggest the structure may be the result of two or three smaller structures that are moving in the 'right direction' simply by 'random chance' which thus came together without gravitational assistance. Others propose that the speed of gravity may have been different in the early universe. But if true, that would open a can of worms in physics.

The LQG also violates the Cosmological Principle which states that the universe is essentially homogeneous when viewed at a sufficiently large scale. This principle is assumed but has never been proven beyond a reasonable doubt. The LQG clump of matter asserts that the universe is not homogeneous.

Dr. Clowes noted, "This is significant not just because of its size but also because it challenges the Cosmological Principle, which has been widely accepted since Einstein. Our team has been looking at similar cases which add further weight to this challenge and we will be continuing to investigate these fascinating phenomena."

Clearly, if the LQG violates the expected speed of gravity and even the Cosmological Principle, something is wrong. Furthermore, it shouldn't even exist—it's too big.

The discovery is a confirmation of the work of astronomer Halton Arp who predicted decades ago that this sort of observation would be made. He should be singled out for recognition; yet, his name is never mentioned in any of the press releases.

Even more startling for modern science is that the LQG cluster shatters another foundation of the Big Bang theory. A structure of such magnitude could not have formed in the time frame of the Big Bang explosion.

When such baffling discoveries are announced it makes us wonder, *do we really know the complete story*?

Dr. Clowes conceded, "So this represents a challenge to our current understanding and now creates a mystery rather than solves one."

Black Holes

All the matter, stars, and gas clouds in the Milky Way revolve around a giant black hole four million times the mass of our Sun. Other galaxies have black holes that extend up to a billion times the mass of our Sun. How was there enough matter in the early cosmos to plummet into these super massive black holes?

There are countless objects that orbit around black holes indefinitely, yet a small sideways velocity prevents them from falling in. This reveals a disturbing problem. If revolving bodies don't fall in then black holes shouldn't be as massive as they are. Physicists have no good theories to explain how black holes became so massive in the early universe.

Do we truly understand black holes? They have never been seen directly because not even light can escape from their gravitational clutches. It means they are invisible and shrouded in mystery. For the time being, black holes are theoretical entities derived via mathematics—not by direct observation.

Pulsars are another puzzle that defies our knowledge of physics. A pulsar is a rotating neutron star emitting a beam of electromagnetic radiation that endlessly pulses in and out. However, 20 percent of these stars pulse wrong according to the predictions of our current knowledge of physics. They have a pulse which can't be explained by any theoretical model. Some physicists postulate that exotic physics might occur with pulsars.

Science has a standard theory how the solar system was formed. A huge number of rocks eventually smashed together to form massive bodies like Earth, Venus, and Mars. That's the conformist theory. But these bodies were originally orbiting within a spinning gas cloud. There would have been friction between the rocks and the gas which should have caused the rocks to spiral into the Sun in a time scale too short for the planets to form.

The orbits of the planets present another mystery. Their

journeys across the skies can be tracked precisely using mathematics even though they just coalesced from gas and dust. Science tells us it came about due to a decline in temperature. But how do clouds of dead, diffuse gas and dust become bodies that revolve and rotate from a drop in temperature? In summary, it is not clear, either from the math or from the models, how planets form, rotate, and revolve in predictable orbits.

Stellar novae provide another paradox. Some physicists say they occur every 30 years or so when a star dies and becomes a super nova. Others estimate the figure to be about 40-60 years. Yet, astronomers have observed less than 300 super nova rings (remnants of explosions) in the entire cosmos. If the universe is billions of years old, modern telescopes should detect millions of such rings.

It seems that the evidence coming from the stars themselves indicates that they might not be billions of years old. Rather, they are "suspiciously young" as originally stated by Edwin Hubble.

BIG BANG OR BIG BLUNDER

Hubble suggested the universe was suspiciously young, and now some scientists are claiming that it's suspiciously small. Mathematician Jeffrey Weeks, and astrophysicist David Spergel, both claim to have evidence to support this hypothesis.

Nature magazine published the following report:

> "Scientists have announced tantalizing hints that the universe is actually relatively small, with a hall-of-mirrors illusion tricking us into thinking that space stretches on forever...Weeks and his colleagues, a team of astrophysicists in France, say the WMAP results suggest that the universe

> is not only small, but that space wraps back on itself in a bizarre way."[113]

However, if the cosmos is flat and infinite, it is indeed bizarre that "space wraps back on itself."

New Scientist published a report by David Spergel on the same topic, wherein he also mentioned the hall-of-mirrors analogy:

> "Effectively, the universe would be like a hall of mirrors, with the wraparound effect producing multiple images of everything inside." Spergel adds: "If we could prove that the universe was finite and small, that would be Earth-shattering. It would really change our view of the universe."[114]

Another issue is that if galaxies are receding from us at enormous speeds determined by the Big Bang model then they should have broken their gravitational bonds long ago. How is gravity still holding these deep-space galaxies under its grip according to modern gravitational theories?

Big Bang cosmology answers the galactic anomaly by the assumptions of dark matter and dark energy. Physicists claim that dark energy is propelling the stars and galaxies apart, while the gravity of dark matter is holding everything together. Keep in mind that dark energy and dark matter are only theoretical constructs; an attempt to explain the unexplainable.

Dark energy is described on the NASA website this way:

> "Something, not like matter and not like ordinary energy, is pushing the galaxies apart. This 'stuff' has been dubbed dark energy, but to give it a name is not to understand it. Whether dark energy is a type of dynamical fluid, heretofore

---
[113] *Nature*, vol. 425, p. 593
[114] *New Scientist*, October 8, 2003

> unknown to physics, or whether it is a property of the vacuum of empty space, or whether it is some modification to general relativity is not yet known."

Dark matter and dark energy remain unobservable, unknowable, and hypothetical. The most powerful instruments today cannot detect anything to substantiate their reality. This being the case, can we base our cosmology on their existence and still call it science?

The Big Bang model seems to be toppling on a precipice. World renowned astronomer Fred Hoyle was never shy to reveal the philosophical foundation for his cosmological views. He was convinced that the Big Bang model was based on unscientific assumptions, and expressed his objection of creation-in-the-past to be against the spirit of scientific inquiry:

> "...especially when taken in conjunction with aesthetic objections to the creation of the universe in the remote past. For it seems against the spirit of scientific inquiry to regard observable effects as arising from 'causes unknown to science' and this in principle is what creation-in-the-past implies."[115]

As physicists became more entangled in the complexities of confirming the Big Bang model, they found it harder to extricate themselves from an uncomfortable situation. It meant that the modern story began to resemble the traditional story in that there was a beginning. They were probably better off with the Steady State theory proposed by Hoyle.

The benefit of Hoyle's model was that it didn't have the process problems of the Big Bang model. Nor did it have origin problems because the model posits no beginning. Of course, in

---

[115] Fred Hoyle, "A New Model for the Expanding Universe", *Royal Astronomical Society*, 108, 1948, p. 372, and in his book, *The Nature of the Universe*, Oxford University Press, 1952

another sense, the no-origin concept is too similar to the traditional creation myth where God is considered eternal and has no beginning.

Either way, the modern story is starting to resemble the traditional story. Like the traditional account, the modern equivalent is a story that is far from proven and is beset with major problems, none of which are acknowledged to the general public.

15
The Jury Deliberates

In this book, you and I have come to know that certain aspects of cosmology and physics do not represent fixed truths. They can be described more accurately as a collection of ideas that best represents what physicists understand at any given time.

You have witnessed the case for the modern creation story as would be argued in a court of law. While the jury is deliberating on the verdict we can briefly summarize the points that were analyzed.

Just as dramatic storytelling is the essence of mythology, in the same way myth is being reborn through popular science books, magazines, YouTube, documentaries, and Wikipedia. Science writers capture our imagination by means of a multiverse containing parallel universes with copies of you and me. Yet the data of true researchers often goes unrecognized.

The multiverse myth adds to the mystique that physicists enjoy today. Going back to ancient soothsayers, mystique has clouded the vision of naïve people who are easily impressed due to their lack of knowledge. Hollywood movies and TV shows portray mythology as if it could be reality. Ironically, when the myth insults our intelligence it becomes more palatable (similar to Hitler's Big Lies).

The jury begins its deliberation by discussing the following point. Best-selling science books and TV documentaries present scientific theories that highly regarded physicists, like Richard Feynman, consider to be nonsense, and Sir Roger Penrose considers to be just a collection of ideas and aspirations. A major consequence is that the border between science fact and science fiction has become blurred. Students and teachers find it increasingly difficult to determine the difference.

It's apparent to the jury that the driving force behind 20th century physics was to establish the Big Bang model as the modern creation story. The unsuspecting public was fed a diet of speculation by the scientific community to save the Big Bang theory from collapsing. At the same time, most scientists relegated the traditional creation story to a bronze-age superstition.

The case for the traditional story is based on a supernatural being who created the universe in seven days. The reference is the Judeo-Christian-Islamic chronicle. That explanation has never been proven by hard data; therefore science considers it to be "blind faith without any evidence."

On the other hand, the case for the modern creation story also depends on an extraordinary beginning: gravity and M-theory causing spontaneous creation of the universe from nothing. Or, an explosion from a singularity filled with matter, energy, gravity, mathematics, and universal laws crushed into a subatomic size, which just came into existence from nowhere and burst out to become our universe.

How did gravity and laws of physics come into being to cause the universe? Or, by what mechanism did everything get compacted into a subatomic size initially? What was the situation before it was compacted? There is no data to substantiate or refute modern hypotheses.

A juror comments that when there's an explosion something gets blown up, or destroyed. "How do we accept that all of creation is the result of a Big Bang explosion, unless the singularity was like a human cell that has the DNA molecule to reproduce another human body? Even more challenging is the idea that everything came from nothing. That's a bit of a stretch for me," he admits.

"Another problem is filling in the gaps between dead diffuse gas clouds in space and fast forward to my Uncle Sidney trying to get a good deal on a three piece suit in New York City. I have reservations buying a theory with such major gaps. Call me naïve but I prefer to believe there's a Supreme Intelligent Being who is the brains behind everything. As mind boggling as that seems it still makes more sense to me than the idea that something exploded from nothing."

Another juror remarks that our concept of time and space begins with the creation of the universe. "If some laws of nature, like gravity, caused the universe, then these laws must have existed prior to time. That leaves us with wholly non-physical laws of nature, outside of time and space, creating the universe. It means that those laws themselves are supernatural."

Overall, the jury acknowledges that a large quantity of imagination has spawned current trends in cosmology. And all we have to balance the books is mathematics on paper combined with dark matter and dark energy, both totally unknown, to the tune of 96 percent. "It means the modern account is in the same category as the traditional story. Therefore, shouldn't it also be considered blind faith without evidence?"

Even renowned mathematicians have testified that mathematics does not always represent reality. It can be configured in various ways to establish whatever a physicist wants to prove. A science article in TIME concluded that "no matter how astrophysicists crunch the numbers, the universe simply doesn't add up."

One juror recalls the argument that the laws of the universe are extremely fine-tuned to support life—the Anthropic Principle. Science concedes that even the slightest adjustment in the constants of any law of nature would produce a universe where life as we know it could never arise.

Physicists call this conundrum the Goldilocks Enigma. But why is it so? Simply because they assume that no intelligence was involved in creation. Of course there is no data to verify or falsify this assumption, so it remains in the realm of conjecture. But if physics allows for higher consciousness in the universe, then fine-tuning is no longer an enigma.

In the RT radio interview from Chapter Three, Dr. Fillipenko spoke candidly about the laws of physics.

> "Why are there any mathematical laws of physics rather than just nothing at all? I don't know whether we will ever understand that. Scientists are only well-aware of four percent of the universe—that is, we understand pretty well the nature of four percent of the universe—the stuff that is made of atoms. Ninety-six percent of the universe is made out of dark matter and dark energy. And although we know they are present we don't know what their detailed properties are or why they are there. Or what exactly is going on."

The fact remains; nobody knows how the laws of nature came into being. Nor can physicists say why the laws can be described via mathematics, nor why they continue to work to produce predictable results after billions of years. Some physicists frankly admit that science cannot say.

There are too many unanswered questions and coincidences about the bio-friendly nature of the universe. A juror asks, "How many 'coincidences' does it take before it's no longer coincidental? Grasping at straws has taken on a new meaning." Other jurors agree.

Quantum Mechanics experiments have revealed that the behavior of one particle implies awareness by another particle —quantum entanglement. But who will agree that particles have consciousness? This is an attribute applied to observers, not particles. Yet, particles do seem to communicate with each other so the symptoms of awareness are exhibited.

If awareness is present in sub-atomic particles, and the universe is comprised of an infinite number of particles, then it follows that an infinite consciousness could be present in the universe.

A major difference between the traditional and scientific versions is that the traditional account includes conscious beings. The scientific account is impersonal with no reference to consciousness or conscious beings.

Every juror agrees that consciousness is the hallmark of life. The evolution of life on Earth is evidence that the creation of the cosmos allowed for that possibility. Whether by chance or by design is beside the point, which is, that consciousness must be included in any reasonable account of our universe.

The "hard facts" might be that universal laws *seem* to serve a purpose—the advent of consciousness, which ·of course, means the advent of life. When we ignore this implication, the question arises about unfettered inquiry in cosmology.

Dr. Hawking never explains how the laws came into existence first so that the universe could spontaneously create itself. He sidestepped the question on the "Larry King Show" and we saw him continue to skirt the issue in his book. He could easily say that science doesn't have an answer yet. Such honesty would be good for his credibility. It's a huge oversight. By omitting important admissions and explanations his credibility is called into question.

When a juror asks whether the universe is flat or curved, another asks if it is four-dimensional—three for space and one for time—or if it has ten or eleven dimensions.

Yet another offers this perspective, "I have a very conservative idea of science and that is data comes first. If your theory doesn't fit the data, it just shows that your theory is inadequate. You don't throw away data because the theory can't handle it."

A question comes up whether the cosmos is infinite or finite. What we can see is only the observable universe. What can we know about far distant galaxies whose light will never reach us because the cosmos is expanding faster than light speed? Will we always be in the dark about these realms?

A female juror points out that it's no longer clear that the universe represents all that exists. "Is it really just part of a patchwork quilt multiverse of parallel universes lying beyond our observable universe, and therefore beyond our ability to prove its existence? To me it sounds like a classic case of over-interpretation of noisy astronomical numbers."

Obviously, our ability to directly detect anything beyond the observable universe will remain a mystery. The jury finds this similarity to the traditional creation story quite provocative.

Some physicists confidently predict that parallel universes might really exist, yet we only understand four percent of our own observable universe. But if there are parallel universes how would we ever detect them since they are in different dimensions?

One juror questions whether this is science fact or science fantasy? Another asks how is this different from religious faith?

Ultimately, there are too many unsolved mysteries in physics to verify the current cosmological explanation beyond a reasonable doubt. What is dark energy? And what is dark matter? Nobody can say. Most jurors agree it's not very encouraging to have 96 percent of the universe in doubt and unproven.

A Situation of Gravity

Science still can't explain how gravity came into being. The most common force in the universe is not entirely understood today. What causes gravity remains a mystery. Did it exist before creation? This is what Stephen Hawking believes. He claims gravity causes universes to arise spontaneously out of nothing. Everyone on the jury wants to know how gravity got there in the first place so that the universe could arise spontaneously.

Moreover, if gravity was the causing agency then the universe didn't spring from nothing. Even more vexing is that no theory of physics can explain what gravitational force consists of. Is it inherent in matter, or made of graviton particles that have never been detected? Science has yet to fully explain the fundamentals of gravity or how it originally came about.

Though physicists admit their theories may be speculative, they still contend we should take them seriously. Brian Greene spoke about his understanding of the multiverse theory in a TED Conference filmed in February 2012.

He began by saying, "...speculative though the idea surely is, I aim to convince you that there is reason for taking it seriously, as it just might be right."

His goal is to convince us to take the multiverse seriously, presumably because he takes it seriously. He intends to present evidence to substantiate the possibility of his contention. Let's have a look at that evidence.

In his own words: "Part one starts back in 1929, when the great astronomer Edwin Hubble realized that the distant galaxies were all rushing away from us, establishing that space itself is stretching, its expanding."

Wait a minute! Although Hubble's initial 1929 paper considered redshift to be a Doppler effect of velocity, it wasn't long before he changed his view. He no longer believed that redshift established the expansion of space. Indeed, Hubble spent his entire life trying to establish that redshift should *not* be used to establish the expanding universe.

In all his later writings over three decades—his book, his papers, and his lectures—he held that redshift likely represents an undiscovered principle of physics. He and his colleague, Milton Humason, studied the cosmos from the world's most powerful telescopes for decades and neither of them accepted redshift as a true indicator of velocity. The trial discussed this argument in depth.

If Brian Greene wants to tow the party line about redshift and an expanding universe, that's his prerogative. But why would he deliberately present Hubble in an inaccurate way? Greene is entitled to his own opinion, but not to his own "facts". Is he ignorant of Hubble's body of work, or is he simply parroting the prevailing cosmology? Or worse, is he intentionally misrepresenting Hubble to convince us of his own view?

Popular scientists still maintain that redshift establishes the Big Bang theory and an expanding universe. But physicists have yet to provide genuine data to support redshift interpretation. There is no ironclad evidence to verify the redshift explanation beyond just an interpretation of apparent velocity.

A growing number of modern physicists believe that redshift signifies an unknown principle of physics that has yet to be discovered. Until actual data resolves all doubts about redshift interpretation, today's belief that the universe is expanding remains hypothetical at best. The proposal is less science and more philosophy.

Science writer John Horgan neatly sums up the issue:

> "Cosmology, in spite of its close conjunction with particle physics, the most painstakingly precise of sciences, is far from being precise itself. That fact has been demonstrated by the persistent inability of astronomers to agree on

a value for the Hubble constant, which is a measure of the size, age, and rate of expansion of the universe.

"To derive the Hubble constant, one must measure the breadth of the red shift of galaxies and their distance from the Earth. The former measurement is straightforward, but the latter is horrendously complicated. Astronomers cannot assume that the apparent brightness of a galaxy is proportional to its distance; the galaxy might be nearby, or it might simply be intrinsically bright…

"The debate over the Hubble constant offers an obvious lesson: even when performing a seemingly straightforward calculation, cosmologists must make various assumptions that can influence their results, they must interpret their data, just as evolutionary biologists and historians do. One should thus take with a large grain of salt any claims based on high precision…

"Our ability to describe the universe with simple, elegant models stems in large part from our lack of data, our ignorance. The more clearly we can see the universe in all its glorious detail, the more difficult it will be for us to explain with a simple theory how it came to be that way. Students of human history are well aware of this paradox, but cosmologists may have a hard time accepting it."[116]

And now Einstein's curved space concept has been called into question by data that suggests the universe is flat. A flat uni-

---
[116] John Horgan, *The End of Science*, New York, Broadway Books, 1996, p. 111

verse implies that it can't curve back on itself but goes on forever. To complicate matters even more, a new breed of physicists and mathematicians suggest a hall of mirrors effect—an effect misleading scientists that the universe is infinite.

Are we finally on the right track with String Theory? It does allow us to answer questions about Quantum Mechanics and General Relativity, but we need eleven dimensions to make it work. Moreover, the strings are so infinitesimal that there is no conceivable way to detect them. How will scientists ever be able to prove or even disprove their existence?

Within cosmological circles, the Standard Model and the Big Bang model are accepted doctrines. Yet both are incomplete and lack important components to establish their veracity. For example, explanations about consciousness and gravity are conspicuously absent from both models.

While the jury remains sequestered, deliberating on the points brought up at trial, outside the court life goes on.

16

The Verdict

A Channel Seven morning news report states that the jury will present their decision when court is called into session today. The media is camped out in front of the justice building once again looking for good interviews.

Using a live satellite feed Channel Seven interviews Australian National University physicist Dr. Paul Frances. After a brief discussion about his views on the points raised at trial the interviewer decides to go for a human interest angle.

"I'd like to digress for a moment to bring up another question. Are we alone in the universe?"

"The universe is so enormous," Dr. Frances replies, "it would make sense that life forms should exist somewhere. But we have zero evidence."

UFOs

In September of 2011, the White House weighed in on the alien issue with the following official statement:

> "The U.S. government has no evidence that any life exists outside our planet, or that an extra-

> terrestrial presence has contacted or engaged any member of the human race. In addition, there is no credible information to suggest that any evidence is being hidden from the public's eye."

The statement doesn't address that the vastness of space suggests the probable existence of alien life.

The fastest spacecraft our current technology can create would take us ten million years to get to the nearest star. Clearly we don't have the resources to contact aliens with our present technology. So that option is beyond our capability.

If life does exist in the universe the only way we would know it is if an advanced life form contacted us. And that could present a problem.

Why would extra-terrestrials contact us? Would they need a planet to live on, presumably if theirs got destroyed or overpopulated, or something similar?

If they were so far advanced beyond us that they could travel to our planet, would they be hostile? We humans are hostile to other living beings on Earth. We confine pigs, cows, and chickens, for our food. We cage monkeys, rats, and mice for medical experiments.

Might an advanced life form see *us* as merely animals?

Cosmology on Trial

> They kill other animals for food, so why shouldn't we do the same?
>
> We have our own problems, don't have time to worry about whether they suffer or not
>
> Don't worry, this one was killed "humanely"
>
> I do feel bad, but I could never give up "my" meat, it's just too tasty
>
> God gave us dominion over them, so that means we can murder and use them as we please.
>
> They're not conscious or aware like us, so it's ok to kill them for food

Credit: anonymous cartoon image

The interview is interrupted with the news that the jury has returned to the courtroom and are ready to announce their verdict.

THE JURY'S DECISION

Considerable chatter has overtaken the courtroom when the judge enters. Everyone rises to acknowledge his presence and retake their seats after he sits down. He directs his attention to the jury.

"Ladies and gentlemen of the jury, after reviewing the arguments presented in this court have you arrived at a decision?"

"Yes, your honor," the foreman of the jury acknowledges. In a quiet and reserved tone he clearly presents the jury's conclusion.

"Astronomers, astrophysicists, and cosmologists, have uncovered too many anomalies for which the Big Bang model of the universe has no answer.

"The scientific understanding of gravity, and how the cosmos came into being, is as uncertain today as ever. There is zero explanation for the origin of the laws of nature that govern matter, energy, and the functioning of the universe.

"To teach that the universe was created spontaneously from nothing when gravity and other laws already existed is incoherent, at best. At worst, it is unsure speculation dressed up as explanations representing great progress.

"Other questions remain unanswered. How did the entire universal matter and energy get crushed into a singularity that would create time and space? How did the singularity come into existence? And where was that singularity located before space existed?

"It's unfortunate that Professor Hawking didn't discuss these obvious correlations. He appeared oblivious to critiques beyond his personal understanding. We will not speculate about his mindset, but it's reasonable to say that people speak according to their capacity to understand.

"When we consider that dark energy and dark matter comprise 96 percent of our universe, it's quite clear that the Big Bang theory presents ideas which cannot be verified by data, or confirmed by any technology known to mankind. This fits the cliché of 'blind faith without evidence' as established by Richard Dawkins.

"As far as the redshift interpretation is concerned, there are too many versions purporting to explain the phenomenon. This is due to the fact that there is no ironclad data to support any one hypothesis over another.

"Some modern explanations for the mysteries of the universe appear to be as indeterminate as the traditional creation myths of the ancients. Parallel universes which can never be perceived or detected are presented as viable theories with no evidence to back up the conjecture.

"If this is the new cosmology, it is now on an equal footing with the traditional story which posits that God cannot be perceived or detected. It means that both accounts of the origin of the universe are based on hypotheses that have never been proven. The obvious conclusion is that the science creation story gives little certainty in an uncertain world.

"The scientific account has not been proven beyond a shadow of a doubt. Too many anomalies remain unanswered leaving a huge gap in explanation unaccounted for.

"Many people today claim they don't believe in God, they believe in science. For the general public, modern science has become something to believe in rather than something which is understood. A belief system has spread worldwide that science understands the nature of reality leaving only the details to be filled in. As far as cosmology is concerned, this trial has refuted that belief.

"A conflict now exists in cosmology whether some theories are a belief system based on a particular world view, or a scientific system of inquiry based on reason, hypothesis, evidence, and collective investigation.

"Although there are numerous physicists doing excellent work in the field, the judgment of this jury is that today's scientific creation story appears to rest on a similar support as the traditional story—no data to support its conclusions."

Outside the courtroom, Channel Seven continues its interview with Australian physicist Dr. Frances. When informed of the jury's decision, he acknowledges that scientists don't have answers to many questions regarding creation.

"But," he says, "We can offer a more refined form of ignorance. We basically don't know and we have no good theory. You might as well ask your priest or rabbi or imam. They'll give as good an idea as cosmologists."

Final Summation

Scientists believe the Big Bang happened. But belief is faith, and faith is religion, and religion breeds dogma.

When we consider that all living beings are conscious, the modern account does not suitably explain the universe. What is consciousness and where does it come from?

The entanglement of particles establishes that some type of communication exists between them. The universe is filled with subatomic particles so if awareness exists between such particles, we are justified to conclude that a universal consciousness could exist. That would imply reality may be a manifestation of some form of consciousness.

My own query is, how does the Big Bang model of the universe fare in terms of the preponderance of anomalies? Well, all the questions in this book need answers before the scientific explanation begins to make sense. Until then, the science story will remain a creation myth.

Because of the authority that scientists hold in society today, their theories imply legitimacy regardless of the true state of affairs. This is what we have discovered on our journey through the universe.

In his 1997 biology textbook, Harvard Professor Emeritus Dr. Ernst Mayr stated that there is confusion regarding how science should be practiced. Modern science gives the impression that it is no longer restricted to what can be known through observation and experimentation.

In a 2003 *New York Times* op-ed piece, "A Brief History of the Multiverse," author and cosmologist Dr. Paul Davies presents various arguments that multiverse theories are non-scientific.

> "For a start, how is the existence of the other universes to be tested? To be sure, all cosmologists accept that there are some regions of the

universe that lie beyond the reach of our telescopes, but somewhere on the slippery slope between that and the idea that there are an infinite number of universes, credibility reaches a limit. As one slips down that slope, more and more must be accepted on faith, and less and less is open to scientific verification.

"Extreme multiverse explanations are therefore reminiscent of theological discussions. Indeed, invoking an infinity of unseen universes to explain the unusual features of the one we do see, is just as *ad hoc* as invoking an unseen Creator. The multiverse theory may be dressed up in scientific language, but in essence it requires the same leap of faith."

In August 2011, *Scientific American* ran an article "Does the Multiverse Really Exist?" by physicist George Ellis, who offered a balanced critique of the scientific philosophy by which multiverse theories are substantiated.

Ellis seems to accept the Level One parallel universe hypothesis, even though it lies far beyond the cosmological horizon. Similarly, the multiverse hypothesis of cosmic inflation exists so far beyond our ability to detect, that it's highly unlikely evidence will ever be discovered. Ellis admits there is little hope that testing will ever be possible.

In spite of this admission, he concedes the theories on which the speculation is based do have scientific merit. Therefore, he accepts the multiverse theory as a "productive research program."

"As skeptical as I am, I think the contemplation of the multiverse is an excellent opportunity to reflect on the nature of science and on the ultimate nature of existence: why we are here..."

> "In looking at this concept, we need an open mind, though not too open. It is a delicate path to tread. Parallel universes may or may not exist; the case is unproved. We are going to have to live with that uncertainty. Nothing is wrong with scientifically based philosophical speculation, which is what multiverse proposals are. But we should name it for what it is."

Ellis argues that "scientifically based philosophical speculation" is tolerable as long as we label it as philosophical speculation. He notes that the lack of empirical testability, or even falsifiability, is not a major concern for many theorists.

"Many physicists who talk about the multiverse," he writes, "especially advocates of the string landscape, do not care much about parallel universes *per se*. For them, objections to the multiverse as a concept are unimportant. Their theories live or die based on internal consistency and, one hopes, eventual laboratory testing."

I think it's clear now that the current understanding of the cosmos is muddy. The Standard Model of physics is supposed to account for all known particles and their interactions. Yet it remains incomplete with real problems that need fixing.

Some researchers comment that speculative ideas such as supersymmetry (which posits that every particle has a heavier twin) and the multiverse (with unlimited parallel universes) need to be dropped from consideration. These models not only border on nonsense, but there is zero evidence they are real. In other words, it's similar to religious doctrine. Of course, other physicists have the opposite opinion.

17
For What It's Worth

This chapter is dedicated to ascertaining whether the latest news from cosmology corroborates or refutes the trial verdict of this book.

For starters we can refer to the article we originally cited at the beginning of the book concerning the crisis in physics that *Scientific American* published in the May/2014 issue.

In their commentary, authors Joe Lykken and Maria Spiropulu focused the spotlight directly on the state of affairs in particle physics, citing a widespread panic.

> "...results from the first run of the LHC [Large Hadron Collider] have ruled out almost all the best-studied versions of supersymmetry. The negative results are beginning to produce if not a full-blown crisis in particle physics, then at least a widespread panic."

Moreover, many troubling observations leave physicists clueless. Theorists working in particle physics admit being in a state of confusion. Despite this, Lykken and Spiropulu assure us, "It is not an exaggeration to say that most of the world's particle physicists believe that supersymmetry must be true." The article underscores the likely debacle:

> "This unshakable fidelity to supersymmetry is widely shared. Particle theorists do admit, however, that the idea of natural supersymmetry is already in trouble and is headed for the dustbin of history unless superpartners are discovered soon..."

It means confidence in supersymmetry is sinking like a stone. The lack of evidence to support the theory implies that particle physicists are relying on a belief system, not on verifiable data. Unfortunately, this belief system is taught as a coherent theory to physics students, while real physicists consider it "headed for the dustbin of history".

The authors doubt that the present approach will shed light on the much sought after union between General Relativity and Quantum Mechanics. Instead they feel it might obscure the prospect more than ever:

> "If this approach is to keep the useful virtual particle effects while avoiding the disastrous ones — a role otherwise played by supersymmetry — we will have to abandon popular speculations about how the laws of physics may become unified at superhigh energies. It also makes the long-sought connection between quantum mechanics and general relativity even more mysterious."

Conversely, we may recall that string theorists, instead of coming to a simpler and more all encompassing grand unified theory, predict 10^{500} different possible laws of physics in a supposed infinite multiverse.

After billions of dollars invested, the LHC has found no evidence of strings, no branes, no WIMPs, no exotic supersymmetry particles, no extra dimensions, no gravitons, and nothing beyond the alleged Higgs boson.

The Standard Model of physics still can't clarify the basic picture of reality, the cause or composition of gravity, dark matter, dark energy, or consciousness.

The relevant question is: Should theoretical physicists carry on speculating or begin searching for an innovative paradigm that can actually provide data, make verifiable predictions, and provide answers to basic challenges?

Lee Smolin writes in his latest book, *Time Reborn: From the Crisis in Physics to the Future of the Universe*, that a new paradigm is desperately needed:

> "We need to make a clean break and embark on a search for a new kind of theory that can be applied to the whole universe—a theory that avoids the confusions and paradoxes, answers the unanswerable questions, and generates genuine physical predictions for cosmological observations."

Smolin candidly admits that he cannot supply a novel theoretical basis for a new direction in physics. "I do not have such a theory, but what I can offer is a set of principles to guide the search for it." His goal, then, is to inspire a new breed of theorists to explore prospects for a fresh approach.

Ripples in Space

Another recent announcement by jubilant scientists: a telescope at the South Pole has detected ripples in space from the very beginning of time. This attracted the attention of *The New York Times*. They published an article on March 24/2014 titled "Ripples from the Big Bang."

The reverberations went far beyond the potential validation of astronomers' most cherished model of the Big Bang.

The ripples detected by the telescope BICEP2, were faint spiral patterns from the polarization of microwave radiation supposedly left over from the Big Bang. They are relics from when energies were a trillion times greater than the Large Hadron Collider could ever produce.

The article investigates the impact this discovery may have for cosmology:

> "These gravitational waves are the long-sought markers for a theory called inflation, the force that put the bang in the Big Bang: an anti-gravitational swelling that began a trillionth of a trillionth of a trillionth of a second after the cosmic clock started ticking.
>
> "Astronomers say they expect to be studying the gravitational waves from mountaintops, balloons, and perhaps satellites for the next 20 years, hoping to gain insight into mysteries like dark energy and dark matter.
>
> "The cosmic Kahuna that now dangles before astronomers and physicists is understanding what caused inflation. What is this stuff that "turns gravity on its head" — as Alan Guth of M.I.T., a founder of inflation theory, has put it — and blew up the universe?
>
> "In the years since, dozens of versions of inflation have been proposed, like chaotic inflation, eternal inflation, slow-roll inflation, hybrid inflation, super-symmetric inflation and natural inflation, based on various kinds of fluctuating hypothetical fields."

According to the press release it certainly looks like this is a breakthrough for cosmology, except for the following paragraph.

> "Assuming they are confirmed (and they have yet to be published in a peer-reviewed journal), the BICEP2 results eliminate most of these versions, including the Higgs, according to the Stanford physicist and inflation theorist Andrei Linde. But the winnowing could go on for decades."

Buried in other paragraphs we find more telltale words that give the game away:

> "If the BICEP2 results are confirmed, and if astronomers agree that the ripples were gravitational waves from inflation…"

Clearly, two important "if" considerations still have to be confirmed before cosmologists can claim success. But they do have high hopes:

> "If the chain of evidence and reasoning holds up, however, the BICEP2 waves do bear witness to the most fervently hoped-for unification of all…"

Again, "if the chain of evidence and reasoning holds up," reveals how conditional these ripples actually are. In spite of this, the media presents the data to the public as if success is already a foregone conclusion.

Mario Livio, currently an astrophysicist at the Space Telescope Science Institute, which operates the Hubble Space Telescope, commented about this new find in a blog post.

> "The recent potential detection of ripples from the Big Bang by the BICEP2 telescope has justifiably generated huge excitement. If confirmed, the ripples represent an imprint on the cosmic microwave background by gravitational waves."

Although the findings remain unconfirmed, Livio is excited about the potential. For decades physicists have been hoping for a Grand Unified Theory combining Einstein's Relativity Theory, which works on the universal scale, with Quantum Mechanics, which works on the subatomic scale.

Livio is optimistic that these "gravitational waves are produced through a quantum process, providing, for the first time (again, if confirmed), evidence that *gravity is governed by quantum mechanics.*" [his emphasis]

He is hopeful that this could be a piece of the puzzle that eventually leads to a breakthrough in the search for a quantum theory of gravity. We previously discussed in Chapter Nine, that there is still no quantum theory to account for gravity.

Ripples in Nature

Sean Carroll, author and theoretical physicist at the California Institute of Technology, also published an article about the ripple discovery on March 23/2014 titled, "When Nature Looks Unnatural."

He reminds us that, "Fields like particle physics and cosmology sometimes include good theories that fit all the data but nevertheless seem unsatisfying to us. The Hot Big Bang model, for example, which posits that the early universe was hot, dense, and rapidly expanding, is an excellent fit to cosmological data. But it starts by assuming that the distribution of matter began in an incredibly smooth configuration, distributed nearly homogeneously through space. That state of affairs appears to be extremely unnatural."

Professor Carroll explains the unnaturalness by this noteworthy insight: "Of all the ways matter could have been distributed, the overwhelming majority are wildly lumpy, with dramatically different densities from place to place. The initial conditions of the universe seem uncanny, or 'finely tuned,' not at all as if they were set at random."

He establishes that physicists are often confronted by theories that appear to fit the available data, yet appear unnatural and thus unsatisfying. Still they accept them under the assumption that it's probably just the way the universe is.

In this regard, he remarks that "last week's announcement of the observation of gravitational waves from the earliest moments of the history of the universe will—if the observation holds up—represent a resounding victory for this kind of approach."

Again, the caveat "if the observation holds up" alerts us to the fact that the media is sensationalizing the observation rather than presenting it for its real worth. We already discussed the theory of cosmic inflation, proposed by physicist Alan Guth, in Chapter Seven. He wanted to present a more natural explanation for why the universe looked the way it does.

"Guth's proposal," Carroll continues, "was that the extremely early universe was dominated for a time by a mysterious form of energy that made it expand at a super-accelerated rate, before that energy later converted into ordinary particles of matter and radiation."

Of course, Carroll admits that nobody really knows what the source of that energy was: "physicists have a number of plausible candidates; in the meantime we simply call it 'the inflaton'. Unlike matter, which tends to clump together under the force of gravity, the inflaton works to stretch out space and make the distribution of energy increasingly smooth. By the time the energy in the inflaton converts into regular particles, we are left with a hot, dense, smooth early universe: exactly what is needed to get the Big Bang model off the ground."

So an explanation is proffered to fit the theory. Even though nothing like the inflaton exists in the universe, one has to assume that it might have existed in the beginning for the theory to continue its tenuous existence.

Carroll candidly clarifies that, "Cosmic inflation is an extraordinary extrapolation. And it was motivated not by any direct contradiction between theory and experiment, but by the simple desire to have a more natural explanation for the conditions of the early universe."

So now we know the reason why modern cosmology accepts inflation as "a starting point for much contemporary theorizing about the beginning of the universe. Cosmologists either work to elaborate the details of the model, or struggle to find a viable alternative."

Truth be told, however, Professor Carroll still has his reservations. "If these observations favoring inflation hold up — a big 'if', of course — it will represent an enormous triumph for reasoning based on the search for naturalness in physical explanations."

In spite of the above, there is one major concern even with the big 'if': "The triumph, unfortunately, is not a completely clean one. If inflation occurs, the conditions we observe in the early universe are completely natural. But is the occurrence of inflation itself completely natural?"

Carroll answers his own rhetorical question by pointing out the upshot: "The original hope was that inflation would naturally arise as the early universe expanded and cooled, or perhaps that it would simply start somewhere (even if not everywhere) as a result of chaotically fluctuating initial conditions. But closer examination reveals that inflation itself requires a very specific starting point — conditions that, one must say, appear to be quite delicately tuned and unnatural."

The conclusion is thus inescapable: "From this perspective, inflation by itself doesn't fully explain the early universe; it simply changes the kind of explanation we are seeking."

Keep in mind that physicists like Alexander Vilenkin and Andrei Linde point out that the process of inflation can go on forever. Thus, they posit that the alleged inflaton energy, which conveniently converts into ordinary particles throughout the entire universe, could convert in some places but not others. This would result in localized Big Bangs, eventually producing an infinite number of universes, or a multiverse.

All this theorizing with no observational data, goes under the name of science, although it's a much closer relative of faith. For example, if fundamental laws obey the principles of quantum mechanics, then the probabilities of many different experimental outcomes is expected. And if inflation begins as part of a quantum ensemble, and if inflation goes on forever, then the probability that it creates an infinite number of individual universes becomes likely.

Professor Carroll acknowledges that if the above logic is "less than perfectly convincing, you are not alone. Not that it is obviously wrong; but it's not obviously right, either."

The reason the multiverse isn't right is because the idea represents a significant shift in the rationality underlying inflation. Again we see that the foundation is philosophical, not scientific. And in this way, it resembles religious philosophy which also has no scientific foundation.

Professor Carroll concludes his discussion by pointing out that we can't solve the problem at this point because, "we simply don't know how to do the math. The multiverse is a provocative scenario, but the specific models that predict it are very tentative, far from the pristine rigor one expects of a mature physical theory."

And now in June, 2014, the experts at the Harvard-Smithsonian Center for Astrophysics, who announced the breakthrough in March, admit they may have got it wrong. The admission was published in the journal *Physical Review Letters*.

The team, led by astrophysicist John Kovac of Harvard, acknowledged their models "are not sufficiently constrained by external public data to exclude the possibility of dust emission bright enough to explain the entire excess signal."

THE HIGGS BOSON

Next, an update on the Tevatron Collider at Fermilab in Illinois: It has been shut down. No graviton was detected. Physicists are still sifting through their data hoping to find something.

When all the attention centered on the Higgs Boson at CERN, Fermilab physicists also turned their attention to the Higgs, hoping it might have appeared somewhere in their experiments.

News of the Higgs Boson was first published worldwide on July 04, 2012. I read all the announcements in the US and the UK media. The Higgs is the hypothetical particle that carries mass, but was promoted as the God Particle. The term comes from a popular 1993 science book that presented an overview of particle physics.

The manuscript was co-written by Nobel Prize winner and Fermilab director Dr. Leon Lederman along with science writer Dick Teresi. It was originally titled *The Goddamn Particle—If the Universe is the Answer, What is the Question?*

That "goddamn particle" was the Higgs Boson. Lederman explained that his publisher objected to the word "goddamn" and suggested the God Particle instead. It sounded much nicer and would definitely increase sales.

Lederman accepted the term God Particle. In his mind the Higgs Boson was extremely elusive and it appeared critical, at the time, to understand the ultimate structure of matter. He renamed the eighth chapter of his book "The God Particle At Last."

When the book was published the Higgs Particle quickly became known as the God Particle. It not only increased attention but also increased sales. A great marketing tool if ever there was one...

Back to the Higgs Boson 2012 media circus. The headlines and the actual facts were like night and day in every report I read. This statement from one press release says it all:

> "Although their results are said to be strong enough to claim an official discovery, the scientists will avoid doing so because they remain unsure whether the particle they have found is indeed the Higgs."

CERN scientists believe they found some particle, but "remain unsure" that it is the Higgs? After spending billions of dollars they couldn't say whether the Higgs particle was factually found?

Of course, unsure isn't science. It's not even faith.

Stephen Hawking weighed in on the issue to affirm the find was of major importance.

> "If the decay and other interactions of this particle are as we expect, it will be strong evidence for the so-called Standard Model of particle physics, the theory that explains all our experiments so far."

Again we see that everything is based on that pesky little "if" word. Apparently, the Higgs Boson is crucial to support the Standard Model of physics. But the model is in dire need of fixing because it doesn't even take gravity into account. We have already discovered that physicists have yet to fully understand gravity—the most common force in the universe.

Apparently, the existence of the Higgs boson will plug a gaping hole in the Standard Model. Yes, until this 2012 announcement there was a gaping hole in the Standard Model of physics! Of course, nobody ever mentioned this before...

Since the 1960s, the elusive Higgs Boson was considered fundamental to the Standard Model. It was like finding Darwin's missing link. A few decades later, in the 1990s, string theory became quite prominent and strings became fundamental. When Peter Higgs theorized his particle in the 1960s, however, string theory was little known.

According to string theory, what physicists previously mistook as particles are today accepted as vibrating strings. These strings are now considered fundamental entities. This means that particles like the Higgs Boson are either strings or no longer fundamental entities, as argued in Chapter Nine.

Incidentally, most physicists hate the term 'God Particle' including Peter Higgs. But the media loves it. Ironically, the existence of a Higgs Boson only applies to the 4 percent part of the universe explained by the Standard Model. We will learn nothing about 96 percent of the cosmos made up of dark matter and dark energy. As a result, the Standard Model is beginning to resemble Swiss cheese—beset with holes.

In an interview with *The New York Times*, Harvard University physicist Lisa Randall explained that, "no one thinks the Higgs is the final word about what underlies the Standard Model of particle physics [the theory that describes the most basic elements of matter and the forces through which they interact]. Even if the Higgs boson is discovered, the question will still remain of why masses are what they are."

Randall emphasized that, "We all expect a richer theory underlying the Standard Model. That's one reason the mass [of the Higgs Boson] matters to us. Some theories only accommodate a particular range of masses. Knowing the mass will give us insight into what that deeper underlying theory is."

And what happens if the Higgs is shown not to exist? What then?

"The great irony is that not finding a Higgs boson would be spectacular from the point of view of particle physics, pointing to something more interesting than the simple Higgs model. Future investigations," Randall insists, "could reveal that the particle playing the role of the Higgs has interactions aside from the ones we know have to be there for particles to acquire mass."

It would be a more spectacular event if the Higgs boson was *not* found? This isn't what the media is promoting. Either way, it looks like a win/win situation for physics.

Could there be another explanation?

"The other possibility," says Randall, "is that the answer is not the simple, fundamental particle that the Large Hadron Collider currently is looking for. It could be a more complicated object or part of a more complex sector that would take longer to find."

I think it's clear that the discovery of the Higgs is not as big a deal for physicists as the media makes out. On the other hand, if the particle doesn't exist it would mean dumping the Standard Model and going back to the drawing board. So some people have a vested interest in the Higgs.

My take on the 2012 Higgs announcement is that CERN scientists are striving to shore up their financial backers. The publicity also educates the uninformed public who don't know a boson from a bassoon. Positive press is what keeps the coffers full.

Then on March 14 2013, CERN issued a second statement:

> "CMS and ATLAS have compared a number of options for the spin-parity of this particle, and these all prefer no spin and positive parity [two fundamental criteria of a Higgs boson consistent with the Standard Model]. This, coupled with the measured interactions of the new particle with other particles, strongly indicates that it is a Higgs boson."

This announcement also makes the particle the first elementary scalar particle to be discovered in nature. How does the public benefit from all the money pouring into CERN? Well, one person who genuinely benefits from the "discovery" is Peter Higgs. He gained celebrity and as luck would have it, he actually did pocket a Nobel Prize and some cash.

As a final comment, did anyone notice the following sentence in the *New York Times* report at the beginning of this chapter? "Assuming they are confirmed (and they have yet to be published in a peer-reviewed journal), the BICEP2 results eliminate most of these versions, including the Higgs, according to the Stanford physicist and inflation theorist Andre Linde."

So the Higgs has now been minimized. The conclusion is that every physicist has his own perspective regardless of what is disseminated by the media.

Historically, it was a Calcutta physicist, Satyendra Nath Bose, who mathematically described the class of particles which now bears his name.

Bose wrote a paper entitled "Planck's Law and the Light-Quantum Hypothesis" wherein he found some anomalies in Nobel Prize winner Max Planck's thesis. He sent that paper to Einstein in the summer of 1924.

In the early 20th century, scientists from India eluded international recognition. Einstein received so many manuscripts, yet something about Bose's paper caught his attention. He appended a note at the end of the paper stating that: "In my opinion Bose's derivation signifies an important advance."

Einstein decided to publish and promote Bose's work. By championing his paper, Einstein saved Bose from obscurity and gave him the credibility he needed. It was like a passport into the international physics community. This led to the discovery of low-energy states of particles in super cooled gases called Bose-Einstein Condensates.

One final point about the Higgs boson. The CERN physicists could have said they found the graviton, another theoretical particle they are looking for. But whether a graviton or a boson, have we really solved the question of mass?

Even with the discovery of the Higgs, Lisa Randall has already informed us that: "...the question will still remain of why masses are what they are."

Publish or Perish

Earlier in Chapter Nine, we quoted John Horgan: "sometimes the clearest science writing is the most dishonest..." And now, a detailed article in *The New York Times* titled "Fraud in the Scientific Literature" was published on October 5, 2012. It began with the following disclosure:

> "A surprising upsurge in the number of scientific papers that have had to be retracted because they were wrong or even fraudulent has journal editors and ethicists wringing their hands."

The article admits that the "retracted papers are a small fraction of the vast flood of research published each year, but they offer a revealing glimpse of the pressures driving many scientists to improper conduct."

Improper conduct refers to falsifying research, data, and results, to arrive at a preconceived conclusion.

The article further revealed that in 2011, the prestigious scientific journal *Nature* discovered more than 300 published retractions every year for the last ten years. That's over 3,000 retractions in just one publication!

> "A new study published in the Proceedings of the National Academy of Sciences, has concluded that the degree of misconduct was even worse than previously thought. The authors analyzed more than 2,000 retracted papers in

the biomedical and life sciences and found that misconduct was the reason for three-quarters of the retractions for which they could determine the cause."

Dr. Ferric C. Fang, professor at the University Of Washington School Of Medicine, became curious how far the rot extended. He teamed up with Dr. Arturo Casadevall of the Albert Einstein College of Medicine in New York to examine the issue.

It didn't take long to reach a troubling conclusion: retractions were rising at an alarming rate. Even worse, however, the retractions were a manifestation of a profound problem that Dr. Fang called "a symptom of a dysfunctional scientific climate."

Dr. Casadevall commented, "This is a tremendous threat." As editor-in-chief of the journal *mBio*, he admitted that science has become a winner-take-all game with perverse incentives that lead scientists to cut corners and, in some cases, commit acts of misconduct.

The two researchers took their assessment to the Committee on Science, Technology, and Law, associated with the National Academy of Sciences. The committee members agreed with their concerns.

"I think this is really coming to a head," said Dr. Roberta B. Ness, dean of the University of Texas School of Public Health. Dr. David Korn of Harvard Medical School agreed, "There are problems all through the system."

The report further disclosed that misconduct has now become global. Retracted papers were written in more than 50 countries, although most of the fraud occurred in the United States, Germany, Japan and China.

The New York Times article cited various theories to explain why retractions and fraud have increased, and concluded, "A darker view suggests that publish-or-perish pressures in the race to be first with a finding and to place it in a prestigious journal, has driven scientists to make sloppy mistakes or even falsify data."

Fraud is precipitated by competition in the workplace and anxiety over finances. No matter what deception they use to fabricate, it's about presenting a false sense of innovation.

Although this article focused on medical science, "publish-or-perish pressures" also exist for physicists and astronomers. Nobel Prize winner Peter Higgs publically stated, "Today I wouldn't get an academic job. It's as simple as that. I don't think I would be regarded as productive enough."

The public has trusted physicists more than other sorts of scientists. But physicists are no more immune to personal motivated agendas than any other professional people. They want to protect their reputation and their theories. Grant money, celebrity, and sales of books become a primary focus to advance in the field. They have enormous vested interests.

Acknowledgements

As briefly described at the outset, I began this investigation into Big Bang cosmology by a chance discovery of the book *The Goldilocks Enigma*. Therefore, I want to recognize the inspiration I received from the author, Dr. Paul Davies. My story begins and ends with insights from Professor Davies.

Of course, I must also give immense credit to my beautiful young wife, Christi, who encouraged me along the way, even when the going got tough, and it looked like we were on a wild goose chase that would never end.

I sent the first draft of the book to my friend Joshua Wulf, who made valuable comments, including that he was amazed that I was a physics buff.

I sent the second draft to physicist Mauricio Garrido who was recommended by a friend. Mauricio guided me to papers and ideas beyond what I had found by my own research. He also clarified certain concepts that really needed amplification.

Although I had a manuscript that was logically and scientifically solid, I still needed an editor to make it understandable to the general public not well versed in cosmology. I was led to Noelene Musumeci, a philosophical editor who reviews all sorts of manuscripts. She told me she knew hardly anything about cosmology, but would give it a read. I responded that she would represent the uninformed public that my book hoped to inform, so her input would be invaluable. And indeed it was.

I now had a very strong document that needed a cover, promotion, and sales. As luck would have it, I bumped into an old friend Alister Taylor who was into book publishing and marketing. He contacted Robert Wintermute who expertly did the cover for the book. Then we developed a marketing strategy, the book became a reality, and now you know the whole story.

About the Author

I was born in London, but when I was five our family sailed away to Canada. My mother was a cultured lady who sent me for violin and music theory lessons when I was only six years old.

As a teenager I began to lose interest in the violin and bought my first guitar. I taught myself how to play and soon switched to a Fender Telecaster. Before long, I had formed my first band and was playing locally. A few years later I was gigging in various rock bands around the country.

Music has always been my first love, and still is.

I didn't start out as a writer, although it was predicted by the mother of my college girlfriend. One Saturday afternoon, she was describing how her father was a missionary in Shanghai before World War II. As a girl in China she had studied palmistry, so she asked to see my hand.

After a few minutes study she pronounced that I had a talent for writing. "One day you will become a writer," she said.

Years passed before that writing talent manifested. But gradually my interest in writing, and especially research, increased. I loved digging behind the scenes to find out what was really happening beneath the headlines. That's how I became an investigative journalist.

You can visit my Author Page: http://www.amazon.com/author/pierrestclair

Now if you really like heartfelt blues/rock music, that's where I feel very much at home. Almost all the music I listened to as a kid was from blues and soul singers, until the Beatles came along. But they also listened to those same soul singers when they were kids!

Connect With Me

If you're reading this, it means you purchased my book and have read it.

My heartfelt thanks for your support.

So what did you think? Let me know what you liked, or what you thought was lacking.

I will personally read and respond to questions and comments on my Facebook page.

Here is the URL:
http://www.facebook.com/cosmology.on.trial

Please send me your feedback because it's an opportunity to make this a better book for future readers:
cosmologycrisis@gmail.com

I look forward to hearing from you.

With gratitude,

Pierre St. Clair

Motivational Videos

I hope you enjoyed my book and learned something valuable. These videos are not about physics, but they have inspired me to reach my full potential, in spite of the cheerless people who look to find faults in others.

You can watch them whenever you want to make a positive change in your life. Not every video will resonate with every person, so watch what inspires you.

If I can motivate you to go beyond what other people say or believe, you can come closer to reaching your full potential. Nobody needs to associate with people that can't uplift or inspire you to bring out your best!

You can also send me an inspiring video.

www.youtube.com/watch?v=JSDoPY9B0wQ
www.youtube.com/watch?v=Cbk980jV7Ao
www.youtube.com/watch?v=g_YZ_PtMkw0

www.youtube.com/watch?v=Sv3xVOs7_No
www.youtube.com/watch?v=YpaSZOq0C3U
www.youtube.com/watch?v=CsgaFKwUA6g

www.youtube.com/watch?v=PT-HBl2TVtI
www.youtube.com/watch?v=qX9FSZJu448
www.youtube.com/watch?v=CvQBUccxBr4

www.youtube.com/watch?v=dePmC_mE-b4